Introduction to Petroleum Exploration For Non-Geologists

Introduction to Petroleum Exploration For Non-Geologists

ROBERT STONELEY

Professor of Petroleum Geology
Royal School of Mines
Imperial College, London

OXFORD   NEW YORK   TOKYO
OXFORD UNIVERSITY PRESS
1995

Oxford University Press, Walton Street, Oxford OX2 6DP

Oxford   New York

Athens   Auckland   Bangkok   Bombay   Karachi
Calcutta   Cape Town   Dar es Salaam   Delhi
Florence   Hong Kong   Istanbul   Karachi
Kuala Lumpur   Madras   Madrid   Melbourne
Mexico City   Nairobi   Paris   Singapore
Taipei   Tokyo   Toronto

and associated companies in
Berlin   Ibadan

Oxford is a trade mark of Oxford University Press

Published in the United States
by Oxford University Press Inc., New York

A catalogue record for this book is available from the British Library

Library of Congress Cataloging in Publication Data
Stoneley, R. (Robert), 1929–
Introduction to petroleum exploration for non-geologists / Robert Stoneley.
Includes bibliographical references (p.   ) and index.
1. Petroleum—Prospecting—Popular works.   2. Petroleum—Geology—
Popular works.   I. Title.
TL271. P4S86   1994   622'. 1828—dc20   94-28504
ISBN 0 19 854856 7

Typeset by EXPO Holdings, Malaysia
Printed in Great Britain by
the Bath Press

Foreword by Dr M.J. Fisher, Chairman of JAPEC

This book has its origins in a 3-day short-course with the same title, organized by the Joint Association for Petroleum Exploration Courses (UK) (JAPEC). Following an initiative by UK-based petroleum exploration companies, JAPEC was formed in 1980 to provide a training service for the petroleum industry under the sponsorship of the Geological Society, the Petroleum Exploration Society of Great Britain, and the Department of Geology at Imperial College in the University of London. As the first Chairman of JAPEC, Professor Stoneley recognized at the outset that the large population of non-technical staff in exploration and production, and service, companies, or in organizations with related interests, were poorly served by existing training programmes. He devised a course to satisfy this need, and in 1981 'An Introduction to Petroleum Exploration for Non-Geologists' became the fifth course to be presented by JAPEC. The course has been presented, by Bob Stoneley and Tony Grindrod (University of Greenwich), initially also with Robert Kinghorn (Imperial College), 15 times in the past 13 years to over 1100 attendees, as well as 'in-house' for a number of companies. The continued popularity of the course is both a vindication of the original perception that the effectiveness of this vital component of the petroleum industry's workforce would be enhanced by appropriate training, and an indication of the excellence of the course itself.

The JAPEC course is supplemented by a set of comprehensive course notes which, over the years, have been in great demand as a reference manual. It was logical therefore to take the next step and prepare the course notes for publication to reach a wider audience. It has not been such a simple task as this bland statement would suggest. Figures that were acceptable in the context of supplementary course notes have been redrafted to publication standard, and the whole text has been rewritten so that it is fully informative and independent of any material contained in the course. However, the successful formula established in the course has been carried through into the book, in the broad scope of topics discussed and in the clarity with which sometimes complex ideas are explained.

Key topics covered in this book are: the underlying rationale for exploration; essential basic geological principles; the nature, formation, and trapping of petroleum; why petroleum occurs where it does; an explanation of geological and geophysical exploration techniques; drilling and logging wells; an outline of reservoir geology and what constitutes reserves; and all of these topics are brought together in a case-history overview of exploration in the North Sea.

The book is fully self-explanatory, needs no prior knowledge of geology or of exploration techniques, and as much as possible avoids the use of technical jargon. There is, however, a lengthy glossary of technical terms that can be used as a reference. The appeal of the book extends far beyond the audience originally envisaged for the JAPEC course. Anyone interested in finding out what petroleum exploration is all about will enjoy this volume.

Acknowledgements

The author is deeply indebted to all those people in JAPEC who, over the years, have encouraged and supported the short-course on which this book is based. In the early days, fellow contributors included Mr Tony Grindrod (now at the University of Greenwich) and Dr Robert Kinghorn (Imperial College): Tony is still a partner in the presentation of the course. The author is particularly grateful to these colleagues for their inspiration and companionship. Thanks are also due to the many people who have attended the course in the past for their interest and encouragement, as well as for constructive comments which have enhanced its format; the course has evolved considerably since the first presentation in 1981.

Appreciation is due to the Department of Geology at Imperial College for encouragement in the production of this book during the author's full-time employment. In particular, considerable assistance has been provided by the late Tony Brown, who drafted the majority of the figures, and by Adolpho Cash, whose photographic skills have contributed to the production of many of the photographs.

Dr G.D. Hobson and Dr M.A. Ala have generously and ably checked drafts of the entire book, and Mr I. Williamson has commented on the section on geophysical techniques. The writer is deeply grateful to these gentlemen for their wisdom and constructive suggestions.

Various members of the Department of Geology have contributed pictures: Prof. H.D. Johnson, Dr J. McM. Moore, Dr M. Rahman, Prof. R.C. Selley, Dr H.F. Shaw. Other photographs were kindly provided by Dr M.W. Hughes Clarke, Prof. A.R. Lord, and Mr A.T. Pink. Companies giving assistance with illustrations have been The British Petroleum Company, Royal Dutch Shell, Schlumberger, and Nopec. To all of them sincere thanks are due. All photographs that are not attributed were taken by the author.

Finally, the patience of Hilda, my wife, and my family in putting up with my preoccupation and irritability is gratefully recognized. Without such support, the book would never have seen the light of day.

Contents

1

Introduction

This book is designed for those interested or involved in petroleum exploration and development, but who are not graduate Earth scientists. It is hoped that it will be useful for draftspeople, secretaries, personnel people, engineers, accountants, and lawyers—to name but a few. Hopefully it may make their work more interesting and effective: it is always desirable to have some idea of what one's colleagues are doing, why, and what their problems and constraints are.

In their daily work, most 'explorationists' (a ghastly but commonly used term—see below) will be spending much of their time shuffling paper, or pushing a mouse to produce coloured images on the screen of their workstation. We are not concerned here, however, with how this day to day work is done; rather we are concerned to give a picture of why they are doing it, and what it is all about. Many of the tools that are used, including interactive workstations, continue to evolve very rapidly. However, most of the basic principles of exploration, and development, have changed little in the last two decades; the answers that we are seeking today with our wonderful new black-boxes, are essentially those that were sought twenty years ago with less sophisticated equipment, and perhaps with less precision. It is the reasons and purposes behind these that this book aims to explain.

The whole subject of exploration and production has now become so wide-ranging and complex, that only the basic principles can be described. Interested readers can pursue their curiosity through some of the additional reading suggested in Appendix A: they should be sufficiently well-informed to be able to use these as a follow-up to this book. To help to try to understand the abundant jargon, some of the common and more important technical terms are set out in a Glossary at the end of the book (Appendix B); they are italicized in the text where they are first used. And, just in case anyone should feel overwhelmed, let him or her take comfort from the fact that the whole business is today so broad and involved that no-one, however

brilliant, can fully master everything. Exploration is carried out by teams of specialists, all with the need to understand each other's problems and to communicate effectively.

To help us on the way, here are three words that are commonly used and misused:

1. *Petroleum* is useful, not because we like long words, but because it is a blanket term to cover all the naturally occurring *hydrocarbons* (a word which itself is sometimes used interchangeably with 'petroleum'): gas, crude oil, and also certain solid substances are embraced. Thus gas is gas, oil is liquid, pitch is pretty solid; 'petroleum' covers the lot.

2. One commonly hears about 'oil deposits' or even 'gas deposits'. This is incorrect. Except in very rare circumstances, petroleum is *not* deposited; it accumulates—in accumulations referred to as *pools* or *fields* (see Section 2.3.1). If you hear people talking about oil deposits, you will know that they do not know what they are talking about!

3. There is no good word to cover all of the specialist Earth scientists who may be involved in exploration and production (hence the use by some of 'explorationist'). To a Greek scholar, the word *'geologist'* would embrace them all; it does, however, carry a certain image, and geophysicists and geochemists, for example, are liable to feel slighted and under-appreciated if so called. At the risk of incurring their displeasure, the term is used here in its widest sense to include them all.

1.1 THE PHILOSOPHY OF EXPLORATION

It is assumed throughout that the object of the exercise is to find oil and/or gas in commercial quantities, and to produce it in such a way as to give maximum return on the capital spent. Although of course the petroleum industry provides a vital social service, in, for example, delivering the wherewithal to run our motor-cars, the ultimate objective of any company must be to ensure a return to shareholders. A company that fails to do so will soon find itself out of business. In work using so much exciting science and technology, we must never lose sight of this fundamental truth.

With this in mind, we can regard exploration as encompassing four aspects:

1. *An understanding of the nature, formation, and occurrence of petroleum.* If we go stalking an animal (with a camera, of course!), then first we must know what it looks like, where it lives, and what its habits are. Similarly, we must know how and under what conditions petroleum is formed, the sort of environment in the subsurface in which it occurs, and as much as we can of its habitat in different regions of the world. This is the objective, scientific study of the geology of petroleum, and the understanding that we can reach is what we shall apply in exploring in new areas.

2. *The use of this knowledge to predict.* We can never know for sure whether or not petroleum is down there under the ground in a particular place, until we have actually spent our money (a lot of money) and drilled a well to look. If we could be sure, then we should never drill a dry hole. Similarly, we cannot know exactly how much we may have discovered in a particular field, until we have produced all that we ever shall produce. We are always dealing with a greater or less degree of uncertainty.

What we have to do then, and it is all that we can possibly do, is to come up with our best prediction that there will be something there, and our best evaluation of how much. It cannot be emphasized too strongly that, in this business, there are no 'right answers' or rather, perhaps, that we cannot know them until it is too late. This is the nature of the games we are playing, and it is the essence of exploration.

We are therefore squarely in the forecasting business, just like the weatherman who even today does not always get it right. Coming up with our best forecasts is what exploration is all about; and the best exploration geologist is the one whose predictions turn out to be correct more often than anyone else's. This is what, in the last resort, we are paid for by our companies: to make our best forecasts about the unknown and unknowable. Exploration has been described as the most expensive gamble in history; that is indeed the truth, and our job is to improve the odds—it can be great fun!

3. *The development of special techniques to assist exploration and production.* In the course of any activity, specialized methods and tools are developed to help collect information, to process it, and to apply it. The petroleum industry has always been particularly inventive and innovative: numerous techniques have been developed, many of them now seeing a much wider application. Our job involves understanding and using them. Some of them have become very complex and are almost specialized industries in themselves: nobody can be fully skilled in them all. Nevertheless, the exploration geologist must have a broad understanding of them, be able to appreciate their applications and limitations, and be able to use their results. Some of the more important ones for exploration purposes are outlined in this book: they include geophysical methods (particularly seismic), the drilling of wells, and the collection of information from them.

4. *The quantitative evaluation of discovered and undiscovered reserves.* Petroleum is of course a resource, and a resource means money. As implied above, we are not interested in finding quantities too small to give us the return to cover our costs; it would be wasting our money to drill if that is all we can reasonably expect. We must therefore try also to predict the quantities involved, to put numbers on our forecasts; whereas the weatherman can get away with forecasting that it will rain heavily, we have to make more precise estimates in terms of so many barrels of oil (1 barrel = 42 US gallons or 35 Imperial gallons) or cubic feet of gas.

To try further to get the enormity of these tasks into perspective, let us remember that Nature herself is almost infinitely variable, some would say perversely so: many things can go wrong in extrapolating from points of information to adjacent areas. Indeed, by its very nature, geology is an observational science: we proceed by making observations and then formulating ideas and hypotheses to explain them. It is usually non-quantitative. In these circumstances, a new piece of information can sometimes force us to change our ideas altogether. It is not uncommon to find 'exceptions' to our cherished theories. Making predictions from them thus becomes doubly uncertain.

It is small wonder then that the geological literature is full of such words as 'probably' or 'frequently': over-simplified generalizations can become misleading—if not downright dangerous. Colleagues must understand this, and hence appreciate why it is notoriously difficult to pin a geologist down to a definite statement—he is only trying to be honest!

So let us now proceed to make a whole series of definite statements. Nearly all of those that we shall make have by now been very well tested and are widely, if not universally, accepted. But ..., we still drill dry holes!

Lastly, a word about units. The petroleum industry mostly developed in the United States, and Imperial units, in some cases an American version of them, were used almost entirely. Later, starting in Europe and spreading internationally, metric units began to be introduced. The result is a mess. Thus we may find well depths measured in metres but well bores and drilling bit sizes in inches, quantities of oil in barrels or metric tonnes, gas in cubic feet or cubic metres, etc. Which units one uses tends to depend on where, and when, one was brought up, and where one has worked. In this book, mostly Imperial units are used, but should we occasionally slip into metric, it is hoped that the reader will forgive us. Some of the more commonly used conversion factors are given in Table 1.1.

1.2 SOME BASIC GEOLOGY

In order to understand the occurrence of oil and gas, it is essential to have at least some knowledge of the basics of certain aspects of geology, which is the study of the rocks our planet is made up of, their history, and relationships. This chapter provides a brief overview of those aspects of it that are particularly relevant to petroleum.

1.2.1 Sedimentary rocks

Petroleum, which is the second most abundant fluid after water in the outer layers of the Earth, is particularly associated with the *sedimentary rocks*. These, as the name implies, are the rocks that were formed from sediment that originally accumulated over vast periods of time on the sea-floor, in lakes, and in certain cases on land.

We are not directly concerned, in petroleum geology, with those rocks that form the greater part of the Earth's crust, either those that have formed by the cooling, crystallization, and solidification of material from the liquid interior, the so-called *igneous rocks* (granites and volcanic lavas, for example), or those that have been reconstituted by pressure and heat from deep in the Earth, the *metamorphic rocks* (gneisses, schists, slates, etc.).

Clastic sedimentary rocks It is the weathered debris from pre-existing rocks, including igneous and metamorphic, that goes to form the sediments in which we are interested. The very slow processes of erosion of upland regions cause pieces to be broken off rock outcrops. These are eventually transported by gravity (in screes), by running water (Fig. 1.1), even by the wind, or by ice, to lowland areas, lakes, or the seas, where they settle out and accumulate to form *clastic sediments*: more properly we should refer to them as *siliciclastic*, since the most abundant component is silica.

As this detritus is transported, the pieces get knocked about, broken up further, and individual fragments have their corners smoothed off. They also get sorted according to size by the varying strengths of current and wave action. By the time the sediment reaches the shore, or is further carried out to sea, it is commonly well enough sorted to be distinguishable as shingle, sand, silt, or mud. Everyone will have noticed the different grades of sediment on today's beaches, especially at low tide (Fig. 1.2).

As time goes on, so more sediment is spread out over what was already there. We can see this if we dig deep down through a beach. We may expect to pass down through layers of slightly different composition

Table 1.1 Some units and conversions commonly used in exploration

Length	1 inch	= 2.54 centimetres
	1 foot	= 0.3048 metres
	1 mile	= 1.6093 kilometres
Area	1 square inch	= 6.4516 square centimetres
	1 acre	= 0.4046 hectares
	1 square mile	= 2.59 square kilometres
	= 640 acres	
Volume	1 cubic foot	= 0.0238 cubic metres
	1 gallon (Imperial)	= 4.5461 litres
	1 gallon (U.S.)	= 3.7854 litres
	1 barrel	= 5.615 cubic feet
	= 42 Gallons (U.S.)	= 0.159 cubic metres
	1 acre-foot	= 7758 barrels
Mass	1 pound = 16 ounces	= 0.4536 kilograms
	1 ton	= 1.01615 tonnes
Pressure	1 pound per square inch (psi)	= 16.0185 kilograms per square metre
	1 atmosphere	= 14.7 psi

Fig. 1.1 The process of erosion illustrated by Ben Eighe in NW Scotland. Pieces of the solid rock form screes and get broken into small fragments, which are transported by streams ultimately to reach the coast.

Fig. 1.2 Sediment reaching the coast is sorted by waves and tides on the beach at Red Point, NW Scotland. The lower beach is clean sand, while the upper beach is partly pebbles (*middle right*) and partly boulders (*near left*).

Fig. 1.3 Successive layers of sand revealed by a stream cut through the beach at Cove, NW Scotland. The lower layers are covered and protected by the higher ones. If this process of burial is continued, the lower beds will be compressed and ultimately hardened to sandstone.

Clastic sediments

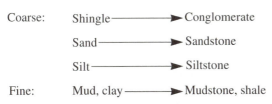

Coarse: Shingle ⟶ Conglomerate

              Sand ⟶ Sandstone

              Silt ⟶ Siltstone

Fine: Mud, clay ⟶ Mudstone, shale

Endemic sediments

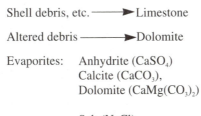

          Shell debris, etc. ⟶ Limestone

          Altered debris ⟶ Dolomite

          Evaporites: Anhydrite ($CaSO_4$)
                           Calcite ($CaCO_3$),
                           Dolomite ($CaMg(CO_3)_2$)

                           Salt (NaCl)
                           Potassium salts

          Plant material ⟶ Coal

Fig. 1.4 The principal categories of sediment and the sedimentary rocks formed from them.

Fig. 1.5 A conglomerate of Triassic age at Budleigh Salterton, southern England. Hard pebbles are set in a matrix of softer sandstone.

Fig. 1.6 Jurassic sandstone near Bridport, southern England. The protruding ledges are formed by harder layers in which the grains are cemented together by lime (calcium carbonate). The sand was deposited originally in a shallow sea.

Fig. 1.7 Jurassic shale in Dorset, southern England. This extremely fine-grained marine mud rock is well layered, but some harder paler beds stand out.

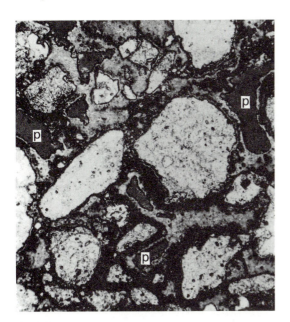

Fig. 1.8 Microscope photograph of a partly cemented sandstone greatly magnified. The sand grains are the pale larger blobs; the cement is pale grey and black. The uniform dark grey areas (p) are empty pore spaces. Photograph from R.C. Selley.

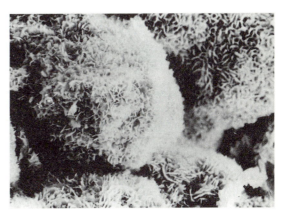

Fig. 1.9 Scanning electron microscope photograph of a sandstone. The large masses are the sand grains, covered almost completely by minute crystals of cement. In the black areas we are looking into the fine pore spaces between the grains. Photograph by H.F. Shaw.

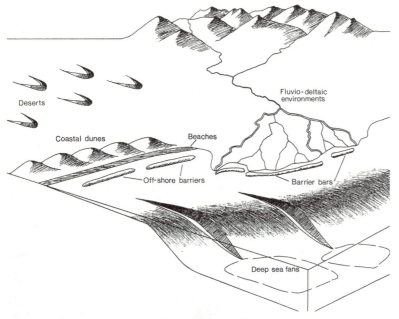

Fig. 1.10 An impression of the geography of a region, as it may have been at a certain time in the past (a palaeogeographical reconstruction). Highlighted are the environments in which sands are deposited, later perhaps to form petroleum reservoirs.

Fig. 1.11 Environmental clues: cross-bedding in Jurassic sandstone in Skye, Scotland. This cross-bedding indicates currents in opposing directions in successive beds. In the bed behind the hammer handle, flow was from right to left, the opposite in the overlying bed.

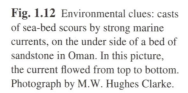

Fig. 1.12 Environmental clues: casts of sea-bed scours by strong marine currents, on the under side of a bed of sandstone in Oman. In this picture, the current flowed from top to bottom. Photograph by M.W. Hughes Clarke.

Fig. 1.13 Environmental clues: ripple marks on the surface of a sandstone bed of Ordovician age in Oman. These are comparable to the ripples commonly seen on beaches today. Photograph by M.W. Hughes Clarke.

Fig. 1.15 Environmental clues: the vertical tubes in the middle of a bed of limestone are infilled burrows made by a type of worm trying to escape from a battering by the waves during deposition of the sediment close offshore. From the Upper Jurassic of Dorset, southern England.

Fig. 1.14 Environmental clues: cracks in the surface of a bed, formed by the sediment drying out in the sun some 200 million years ago. Near Lyme Regis, southern England.

Fig. 1.16 A seam of coal in northern Alaska. It is overlain, behind the author's head, by a thin bed of mudstone, and sandstones form the cliff above. All rocks are Cretaceous in age.

Fig. 1.17 The layering of successive beds is well shown by the Grand Canyon in Arizona, cut through flat-lying Palaeozoic sedimentary rocks. Hard rocks form the cliffs and steps, softer beds the more gentle slopes. An unconformity can be seen in the inner canyon at middle right, and basement metamorphic rocks at the bottom (middle left).

or size grade (Fig. 1.3). The lower layers are also slowly being loaded by the weight of the younger ones on top, so that the water starts to get squeezed out from between the grains, and the sediment gradually becomes harder. Such layers of sediment are known in the jargon as *beds* or *strata*.

This gradual burial is the process which, if it is continued long enough, eventually indurates (hardens) the sediment into a coherent rock (Fig. 1.4). Our shingle becomes a *conglomerate* (Fig. 1.5), sand becomes *sandstone* (Fig. 1.6), silt *siltstone*, and mud hardens into *mudstone* or *shale* (Fig. 1.7). These terms do, for scientific purposes, have strict definitions based on the size of their constituents, but here it should be clear enough what we mean. In such rocks, we can still recognize the original grain components, although we may need a magnifying glass (hand lens) or a microscope to do so (Figs 1.8 and 1.9): these are common geological tools for learning more about the rocks, either in the field or later in the comfort of the laboratory.

Ancient geographies Just as today there are different grades of sediment on different parts of the sea-floor and along the coasts, shingle in one place, sand in another, and mud in a third, so also the rocks in a layer representing any one particular time in the past may show considerable variation when traced laterally across country. If we can work out the environments the sediments were deposited in, and map out their distributions, then eventually we shall begin to see a pattern emerging, a picture of the geography as it was at that time, a *palaeogeography* (e.g. Fig. 1.10). This, if we can make sense of it, is one of the prime tools used in trying to understand (predict) the distribution of different rock types where they have been buried deep in the subsurface and where we can no longer see them.

It is a fascinating piece of detective work to put together such pictures of the environment at different times in the past, using all the clues that we can muster. The kinds of clue that help to indicate the environments of deposition of sediments include: the nature of the rocks themselves; the relationships of one layer, or *bed*, to others; cross-cutting marks within certain sandstone beds (known to the geologist as *current* or *cross-bedding* (Fig. 1.11)), which were caused by running water and which indicate the direction the current was flowing in; scours of the sea-floor, usually preserved as casts on the bottom of some sandstone beds, which can also reveal current directions (Fig. 1.12); ancient beach ripples on sand surfaces

(Fig. 1.13); cracks on the surface of a bed which dried out in the sun (Fig. 1.14); and the types of animal that lived there, either preserved as fossils (see below) or represented by characteristic tracks or burrows that they left behind (Fig. 1.15). Features such as these all provide pointers to the environments and conditions under which the sediment was deposited.

If we have enough information, gradually we can add these clues together to build up a picture of, perhaps, former rivers; estuaries; deltas with swamp forests, the wood now compressed, hardened, and altered to *coal* (Fig. 1.16); beaches; offshore sandbanks; shallow shelf seas; deep-sea canyons, channels, and submarine fans; and so on. We may be able to recognize ancient deserts, with their dunes, wadis, and temporary lakes still discernible. If we glance again at Fig. 1.10 and stop to think about it, we shall realize that most of these environments have a particular place and distribution within the overall pattern. Understanding these provides the basis for our reconstruction of the palaeogeography.

The geography that we draw up for any one time can be expected gradually to be altered, as a result of slow environmental changes over the millions of years; the nature of the sediment deposited at any one place will therefore change in consequence. The relative positions of seas and land masses will alter, so that regions formerly covered by the sea may become dry land, and vice versa (the processes causing these changes are considered in Section 2.4.1). For example, it was only a few thousand years ago that Britain was joined to France across the Channel, much of northern Europe was covered by ice, and the sea-level was appreciably lower than it is today. As a result of global warming (purely natural and nothing to do with man's activities), we now find sand accumulating on the sea-floor off the Norfolk coast whereas, during the ice-age, the deposits of glacial moraines had been left there. The point is that, through time as represented by successive layers of sediment (strata or beds), the nature of the sediments can be expected to alter: we may find a bed of sandstone overlain by mudstone, or vice versa. The picture is one of continual change, which can lead to a succession of very different rock layers on top of one another (Fig. 1.17).

We have now added another dimension to the understanding that we have to reach: that of time, as represented by the successive layers of rock. This study, of the rocks themselves, their age, layering, and interrelationships, is a major branch of geology known as *stratigraphy*.

Fig. 1.18 Oligocene corals in Oman. The growth patterns can be seen on some of the coral heads.

Fig. 1.19 An ancient coral reef of Oligocene age in Oman (detail is shown in Fig. 1.18). The reef was re-exposed when the softer muds that drowned it were later eroded away.

Fig. 1.20 The tomb of Queen Hapshetsut (died 1482 BC) near Luxor, Egypt. The cliff behind is formed by a thick, massive, and well-cemented Eocene limestone overlying mudstones and thin limestones.

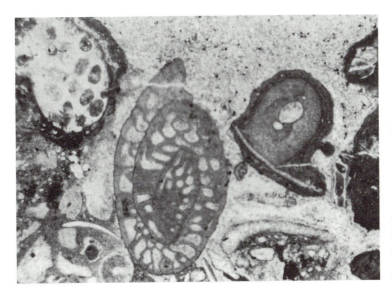

Fig. 1.21 Microscope section through a limestone. Sand grain-size organisms are bound together by stippled calcite cement. Photograph by M.W. Hughes Clarke.

Fig. 1.22 The top of a salt plug in southern Iran. The white salt on the left has punched its way up to surface from a depth of at least 15 000 feet, through overlying sediments, some of which can be seen dipping in the background.

Why all this is important is because each of the various environments of deposition has typical associations of sediment, and because each type of sediment has a characteristic distribution. For example, sand that originally accumulated in a river channel may be expected to extend in a direction roughly at right-angles to a contemporary beach sand. Being able to recognize it as representing a particular environment will help us in predicting the direction in which a body of sandstone in the subsurface may extend. If we can draw up such snapshots of our area of interest as it developed through the whole of its long history, we shall be in a position to map out the distribution of the various rock types associated with petroleum. This is what we are aiming to achieve.

Other types of sediment   The materials forming the siliciclastic sediments that we have been considering were derived from *outside* the area of sedimentation. There are, however, two further important types of sediment, of which the components are endemic to the marine environment in which they were deposited and have not been transported significant distances. These sediments also will of course be incorporated into our palaeogeographies.

An abundance of lime (calcium carbonate, $CaCO_3$) secreting organisms, including certain algae, shell-bearing animals, and corals, can give rise to *limestone*. As their soft parts decay, the harder skeletons of these animals may be preserved, either virtually unchanged (Figs 1.18 and 1.19) or else smashed into pieces by waves or strong currents: whole or broken shells are a familiar sight on our beaches. If there are enough of them, the remains may be transported and eventually deposited as sediment in the same sorts of ways that the siliciclastic sediments are. Once they are buried, squashed, and perhaps cemented together by further calcium carbonate precipitated from the sea-water, we are left with a hard rock composed almost entirely of the mineral *calcite* ($CaCO_3$) (Figs 1.20 and 1.21). A major difference from the siliciclastic rocks is that calcite, unlike silica, is very prone to being dissolved or reprecipitated by percolating waters of varying acidity or alkalinity. We are all familiar with such cementing material as the scale which forms on the inside of a kettle; there of course it is precipitated much more quickly as a result of boiling the water. The result of these processes may be that the entire body of the rock is reconstituted; it may become difficult, or even impossible, to recognize the original shell material, but it is still a limestone.

An important variety is *dolomite*. Particularly in shallow seas off arid tropical coasts, the sea-water may be enriched in dissolved magnesium salts. If this water then percolates through shell debris, the calcium carbonate may be converted, not to another form of calcite as described above, but to a calcium–magnesium carbonate ($CaMg(CO_3)_2$) mineral with the same name as the rock it forms—dolomite. Again, the whole body of a layer of sediment may be affected. We will return to its importance later.

Collectively, limestones and dolomites are commonly lumped together in the useful term *carbonates*.

The second endemic rock type is perhaps even more astonishing. In a restricted or isolated sea or lagoon subject to intense evaporation in low latitudes, the sea salts may be concentrated enough to be precipitated out and to accumulate as solids on the sea-floor. Layers of them may reach considerable thicknesses of hundreds of feet, provided that the sea-water is replenished as it is evaporated. These salts include *anhydrite* or *gypsum* (calcium sulphate), the calcium carbonate that we have already met, *halite* (sodium chloride—common or garden rock salt, ancient deposits of which in fact provide some of the salt that we buy in supermarkets or garden centres), and potassium salts. These salts are collectively referred to as *evaporites*. They are not seen commonly at the surface of the Earth, since they tend to be dissolved by rain-water, but they can be found in some desert areas (Fig. 1.22) and are known in the subsurface in many parts of the world.

These, then, are the principal types of sedimentary rock that we shall encounter in our review of petroleum. But first we must look at some other concepts.

1.2.2 Geological time and the dating of rocks

The processes of erosion, transportation of weathering products, and accumulation of sediment that we have been discussing, take place extremely slowly. Indeed they are virtually unnoticeable in terms of our everyday experience. We are, however, considering their continuation over tens, or even hundreds, of millions of years. It is this element of almost unlimited time that is essential to the concepts of geology.

Geologists talk quite happily about sequences of sediments kilometres thick; they will show no concern at the idea of geography changing even to the extent of new mountain ranges being created and eroded away; they consider the evolution of animals over thousands of generations; and they will envisage petroleum being generated and moving around in the subsurface. In order to appreciate what goes on in their minds, we must remember that all of these

Fig. 1.23 Fossil fish of Eocene age from Colorado. Such a find is very rare in most parts of the world. Photograph by A.C. Cash.

Fig. 1.24 A trilobite from South America, belonging to a family (Dalmanitidae) that lived during the Ordovician and Silurian. Such creepy-crawlies became totally extinct at the end of the Permian. Photograph by A.C. Cash.

Fig. 1.25 An extinct form of sea-urchin (*Micraster*) from the Cretaceous Chalk of southern England. Photograph by A.C. Cash.

Fig. 1.26 A group of fossil scallops and other shells from the Lower Jurassic of eastern England. Photograph by A.C. Cash.

Fig. 1.27 Fern-like fronds from the Upper Carboniferous Coal Measures of England. Photograph by A.C. Cash.

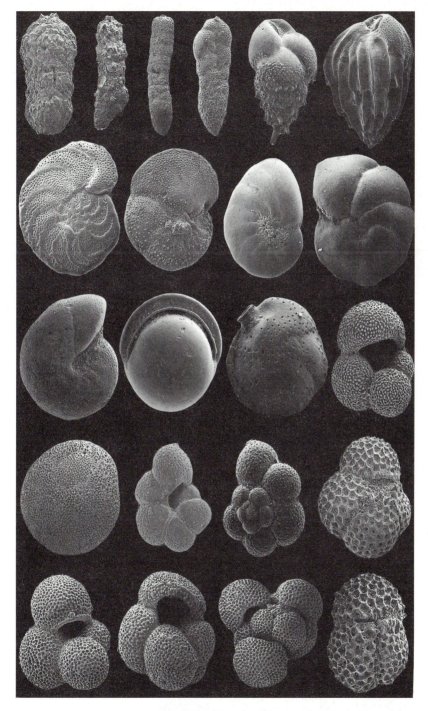

Fig. 1.28 Magnified assorted foraminifera from the Pliocene of Cyprus; these single-celled organisms are about the size of sand grains. The top 2 ¾ rows are animals that lived on the sea-floor, the lower ones are free floating species. Such fossils are particularly useful in oilfield work, since a large number of them may be recovered from a single drill cutting. Photograph by A.R. Lord.

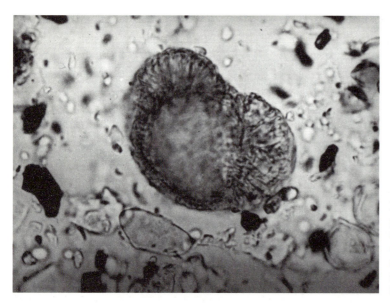

Fig. 1.29 A high-power microscope photograph of a pollen grain. These evolved and changed form just as other organisms and, being extremely small, are very useful in the oil industry. Photograph by Royal Dutch Shell.

processes have taken place over lengths of time that are, in terms of our normal comprehension, almost infinite. Perhaps we can try to get the notion into perspective in two ways:

1. The mind would probably not boggle at the thought of accumulating sediment at the rate of 0.1 mm a year. If it goes on at the same rate, this amounts to 100 metres in 1 million years, or a kilometre in 10 million. It is the sort of rate that we have to account for.

2. The age of the Earth is approximately 4.6 thousand million (10^9 or American billion) years. If we compare this length of time with a 24-hour clock running from midnight to midnight, then the main branches of the animal kingdom (vertebrates excepted) appeared at about 8.00 p.m. in the evening, representing an age of nearly 600 million years. Man evolved about 3 million years ago (depending on what we recognize as Man)—about 2 minutes before midnight—and the Christian era would take no longer than the flick of a finger. A human life-span would be totally insignificant. Perhaps this thought should put us in our places!

The dating of rocks.   Only since the 1960s has it been possible actually to measure the ages of rocks. The process employs the decay times of certain radioactive elements. For example, one method measures the proportion of argon derived from the breakdown of radioactive potassium: by knowing the rate of this breakdown and measuring the amounts present, the age of the rock can be worked out. Other elements are also used, the radiocarbon dating of the archaeologists being a more widely known example. However, these techniques are expensive, and not all rocks contain suitable radioactive materials; it is only by comparing the relationships of those that do with those that do not, that scientists are gradually homing in on ages in millions of years for our sedimentary rocks.

Before these techniques were available, and still today, ages were and are normally expressed in terms of named intervals of relative time (see below), based on the relationships of layers of sediment one to another. There are two fundamental principles involved: firstly that a particular layer is younger than the one beneath it and older than the one on top: and secondly that a bed can be identified by characteristic *fossils* that it contains. These are the remains of animals or plants that lived at the time of deposition and were preserved in the sediment (Figs 1.23–1.29). They are often very beautiful objects, even to the extent of commanding ridiculous prices in fossil shops and jewellers.

Animals and plants have evolved over geological time. Species die out and new ones may replace them; most children know that the dinosaurs became extinct

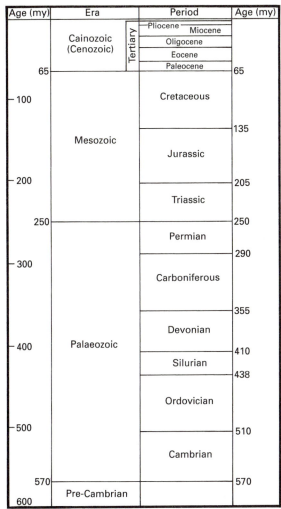

Age (my)	Era		Period	Age (my)
	Cainozoic (Cenozoic)	Tertiary	Pliocene — Miocene	
			Oligocene	
			Eocene	
65			Paleocene	65
100	Mesozoic		Cretaceous	
				135
			Jurassic	
200				205
			Triassic	
250				250
			Permian	
300				290
	Palaeozoic		Carboniferous	
				355
			Devonian	
400				410
			Silurian	
				438
			Ordovician	
500				510
			Cambrian	
570				570
600	Pre-Cambrian			

Age of earth - c. 4600 my

Fig. 1.30 The geological time-scale. Note that the greater part of the Earth's history, before the appearance of most branches of the animal kingdom and therefore of fossiliferous rocks, is not shown.

some 65 million years ago. Experience over more than 200 years of study has enabled specialists (*palaeontologists*) to recognize many stages of this evolutionary process in all branches of the animal and plant kingdoms. If now we are able to identify the particular fossil species that it contains, then we can assign a rock layer to its place in the complete sequence of strata. In the same way, we may be able to identify an individual bed in widely separated localities. As a

general statement, rocks of the same age contain the same fossils. This may apply over a limited area, but it is usually not quite so easy as this. Animals tend to live in particular environments and in particular parts of the world; for this reason, rocks of the same age but in different places do sometimes yield quite different faunas, and we may experience considerable difficulty in correlating even richly fossiliferous sequences of sediments. Nevertheless, fossils still provide the most useful, and widely used, means of relative dating and correlating sedimentary sequences. They have enabled geologists to build up reference sequences of the rocks that represent the time since animal life first appeared on Earth, and their study gives us a very powerful tool (*palaeontology*) for our work. Commonly in petroleum exploration, we are using microscopic animal fossils (*micropalaeontology*) or the spores and pollen from plants (*palynology*).

The rocks that were formed before abundant organized life appeared, the Pre-Cambrian, are likely to be extremely difficult to date and correlate as they contain no fossils. Fortunately, from the point of view of petroleum, this does not usually matter too much because, as we shall see, oil and gas have been produced from the remains of living organisms and are seldom associated with Pre-Cambrian sediments.

The time that has elapsed since the remarkable appearance of most animal groups, some 570 million years ago, has been divided into three eras: the Palaeozoic (early time), Mesozoic (middle time), and the Cainozoic (recent time). They in turn are subdivided into named periods (Fig. 1.30) and even smaller units. These names are what the geologist almost always uses in referring to the ages of rocks: he is not normally programmed to think in terms of numbers of millions of years. Indeed, the absolute time scale is relatively new and is still being refined. The period names are conveniently used also to refer to the rocks that were formed during these intervals of time. Thus we talk about Cretaceous time (approximately 136 to 65 million years before the present) as well as Cretaceous rocks. Until such terms become familiar, the reader may find it useful to flag and refer back to Fig. 1.30.

1.2.3 The evolution of a sedimentary basin

Sediment accumulating in a shallow sea, or on the continental shelf (water depths less than 100 fathoms, 600 feet, or 200 metres approximately), even at the rate that we have envisaged of 0.1 mm/year, would relatively quickly build up to sea-level, where no more could be accommodated. Yet we know of sedimentary sequences that are many thousands of feet thick. In some instances, we can account for this by imagining a pile of sediment building out into the deeper waters of the ocean beyond the continental shelf. However, this cannot have happened, for example, in the Middle East, where there is a huge thickness of sediment that was nearly all deposited in shallow water above the edge of the continent itself. There can be no escape from the conclusion that the sea-floor, and hence an extensive area of the Earth's surface, has been subsiding continuously or intermittently for at least 250 million years.

Regions where such subsidence has taken place, and thick sediments accumulated, are referred to as *sedimentary basins*. They are not necessarily shaped like wash-basins, and indeed the definition is extended to cover also the situation where the sediments build out into the ocean beyond the continental shelf. It is to such sedimentary basins that the search for petroleum must turn: we need the sediments to produce and to house the oil and gas. Let us therefore follow the story of a hypothetical basin, and note some of the astonishing features that may develop (Fig. 1.31). We will try to find how and why such things happen in Section 2.4.1.

When an area of continent first starts to subside, it may well be isolated from the sea. If so, then the first sediments to be deposited, those found at the base of the succession of beds filling the basin, may well have accumulated on land, possibly in deserts, or in river channels or lakes. Sooner or later, however, the sea will break in and a sequence of marine deposits starts to build up. If the rate at which they accumulate keeps pace with the rate of subsidence, then the sea will remain shallow and the sediments will be characteristic of shallow waters, possibly even with temporary periods of emergence. In the opposite situation, on the other hand, if subsidence is faster than the accumulation of sediment, then the water depth will increase and, in general, we can expect to see finer-grained rocks formed, such as muds which later become shales. The difference may be due to many factors, including the supply of sediment eroded from surrounding land areas, the efficiency of its distribution, the climate, and so on.

The processes of subsidence and accumulation of sediment may continue over tens, or even hundreds, of millions of years. As they do so, the lower, earlier layers will become squashed by the weight of the younger overlying ones, any remaining water will be

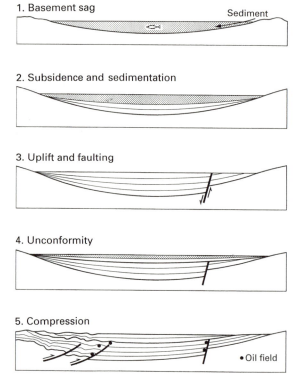

Fig. 1.31 Stages of development and evolution of a hypothetical sedimentary basin. Stage 1: a sector of the Earth's crust starts to subside; sediment derived from the margins may initially be deposited on land or in lakes, but then the sea may break in. Stage 2: marine sedimentation continues, being thickest in the deeper parts of the basin. Stage 3: uplift, perhaps accompanied by faulting and followed by erosion, may affect part or all of the basin. Stage 4: renewed subsidence and transgression of the sea cause further sedimentation, unconformably above eroded older beds. Stage 5: part or all of the basin may get caught up in compression of the Earth's crust; the beds are folded and reverse faulted, possibly forming a range of mountains. Please see text for full explanation.

squeezed out, and gradually they will be converted from soft loose sediment into ever hardening rock, a process often referred to as *induration*. At the same time, by being buried ever deeper into the Earth, they are also gently heated: we will all have seen pictures of miners sweating away as they work deep underground and we will know that it gets hotter and hotter down towards the bottom of the deep mine, down towards the centre of the Earth These two effects, *compaction* and *heating*, cause progressive changes to the rocks that we shall see can be extremely important from the point of view of petroleum.

Subsidence is seldom uniform across an entire basin, and thus relative highs and lows are developed in the basement surface (the very useful term *basement* refers to whatever is beneath the layers of sediment that we are interested in; commonly this is igneous or metamorphic rocks). These may cause local variations in the nature of the sediment being deposited at the sea-floor at any one time: somewhat thicker and perhaps deeper water types in the lows, as compared with the less actively subsiding relative highs.

Subsidence, furthermore, may not go on all the time at an even rate; it can take place intermittently or in jerks. If it stops altogether for a while, the sediment will build up towards sea-level so that no more can accumulate there, and we may thus find that there is a break in the continuity of the sedimentary sequence—a *hiatus*.

During such a hiatus, it may happen that the sediments already accumulated are raised up above sea-level, where they themselves may be exposed to erosion. The beds may become tilted in the process, with one area uplifted more than another and more material being eroded from there as a result. If erosion continues like this for long enough, it will bevel off a surface that cuts obliquely across the bedding of the sediments. Stretch your imagination further now, and picture subsidence and sedimentation starting up again as the sea returns. A new sequence of strata will be deposited horizontally over the tilted and truncated older sequence. The surface representing the interruption in sedimentation and differential erosion is known as an *unconformity*: the term is often qualified according to the more detailed relationships between the two sequences (Fig. 1.32), but this need not worry us here. Surprisingly perhaps, unconformities are not uncommon; examples can been seen in a number of cliffs around our coasts, where once again the sediments have been uplifted, eroded, and exposed to our view (Figs 1.33 and 1.34).

In addition, during the hiatus, or even while sedimentation is continuing, differential subsidence can so stress the sedimentary sequence that it gets broken. Such a break is termed a *fault*, the strata on one side being dropped down or jacked up in relation to those on the other. Fault planes are commonly steep, within say 30° of the vertical, but all angles are possible. An important distinction is between *normal faults*, in

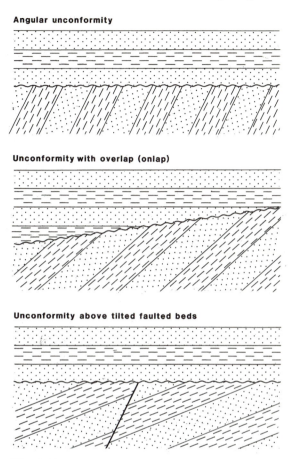

Angular unconformity

Unconformity with overlap (onlap)

Unconformity above tilted faulted beds

Fig. 1.32 Diagrams illustrating possible effects of unconformities. Note that, in the second diagram, the sea will gradually have spread over the eroded surface, extending the area of sediment deposition.

which the side that is dropped down (the *downthrown side*) is in the direction of slope of the fault plane, and *reverse faults*, in which the side that is pushed up (the *upthrown side*) is in its direction of slope (Figs 1.35–1.37): the former is due to a horizontal stretching of the Earth's crust, and the latter a squashing or compression; we shall come to the causes of such stretching and compression in Section 2.4.1. Meanwhile, let us emphasize the difference by looking again carefully at Fig. 1.35. If a well were drilled vertically to cut through a non-vertical fault, it would find that some of the succession of strata was cut out in the case of a normal fault, but that some beds were repeated and drilled through twice if the fault were reverse. This, indeed, is often how we recognize the difference.

Fig. 1.33 An unconformity near Paignton, south-west England. The Permian beds in the upper part of the cliff dip to the right at about 10°; the Devonian strata below dip at some 30°.

Fig. 1.34 An uncomformity in Oman. Permian limestones above the break rest on Pre-Cambrian carbonates beneath. Photograph by M.W. Hughes Clarke.

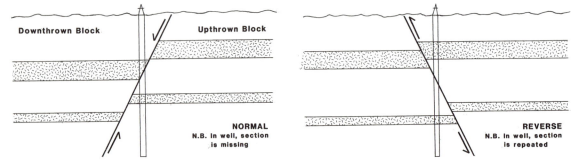

Fig. 1.35 Diagrammatic cross-sections to illustrate normal and reverse faults. Note that a well drilled down through a fault will show a part of the succession of strata apparently cut out if the fault is normal, and will be penetrated twice if it is reverse. Normal faults commonly result from horizontal stretching, so that one side drops down in relation to the other; in reverse faults, usually caused by compression, one side is pushed up and over the other.

Fig. 1.36 A small normal fault in Jurassic beds near Bridport, southern England. The hard bed above the geologist is dropped down some 3 feet to the right.

Fig. 1.37 Steeply dipping hard sandstone and softer mudstone beds of Miocene age near Gisborne, New Zealand. Before being tilted to some 70°, the beds were broken by a fault which is now nearly horizontal and cuts gently up to the right across the top of the picture.

Anticline

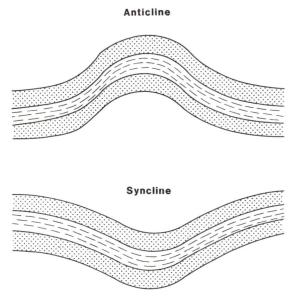

Syncline

Fig. 1.38 Diagrammatic representations of an anticline (upfold) and a syncline (downfold).

Fig. 1.43 The effects of very severe compression shown by Meall a'Ghuibhais, NW Scotland. The rocks forming the upper part of the mountain are *older* than the paler coloured beds running round the base, and have been pushed up and over them.

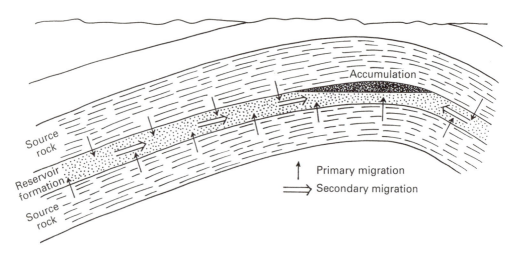

Fig. 1.44 A summary overview of petroleum geology. Petroleum is formed from organic remains, preserved in a source-rock, and gently heated. It is squeezed out and into a reservoir formation (primary migration) through which it is able to percolate upwards (secondary migration) until it reaches a trap where it accumulates; in this case, the trap is at the top of an anticline.

A final thing that can happen to our sedimentary basin is that, sooner or later, part of it may be caught up in the inexorable forces that compress certain zones of the Earth's crust. If this happens, then the originally flat layers of sedimentary rock, as well as being reverse faulted, are likely to be buckled and 'concertinaed' into a series of upfolds and downfolds: an upfold is known as an *anticline* and a downfold as a *syncline* (Figs 1.38–1.40). The more intense the compression, the sharper the folds will become (Fig. 1.41), until they are sheared off (Fig. 1.42) and piled up to form a range of mountains (Fig. 1.43). The very spectacular structures that are visible in the Alps or in the Canadian Rocky Mountains, for example, make it hard to believe that the rocks originally formed as horizontal layers of loose sediment on the sea-floor. If we want to convince ourselves that this did indeed happen, we may be able to go and find the remains of marine animals, or fossils, in the rocks of a high mountain ridge exposed to erosion!

Lastly and before leaving our basin altogether, we should note that very gentle anticlines and synclines can develop during its subsidence. For example, we have already remarked that some parts of it may subside faster than others: a slowly subsiding area will remain relatively high compared with its surroundings, and in effect a broad and gentle anticline will have been created. We do not necessarily have to envisage compression to give an arched form to the rocks; we shall encounter examples in Section 2.3.1.

1.3 PETROLEUM GEOLOGY— AN OVERVIEW

If the reader has persisted and stayed with us so far, he (or she) will have covered a lot of the science of geology, and she (or he) will be able to understand much of what exploration is all about. How, then, does all this relate to petroleum? This is the story that the next few chapters will develop, but a brief summary here may help to keep the whole thing in perspective.

Oil and gas are formed from the soft parts of microscopic organisms preserved in certain sediments, and gradually and gently cooked (*matured*) by exposure to the Earth's interior heat during deeper and deeper burial. It is mainly the remains of marine plankton that become converted to oil and gas, although land-derived plant material (including coal) can generate pure natural gas (methane). To preserve this organic matter from early destruction requires its being interred in an oxygen-free environment, and this is generally achieved where the sediment is an impervious clay. Mudstones and shales are thus the most common *source rocks* for petroleum. Eventually at considerable depths, oil and gas are formed and then slowly squeezed out of the source rock (Fig. 1.44); if they can, they will escape up to the surface of the Earth where they are lost for ever. This escape will take place by percolation along the easiest path available, generally provided by the minute pore spaces between the grains of a porous and permeable rock layer, such as a sandstone or a limestone. These rocks are referred to as *reservoir rocks*. The migrating petroleum, if it reaches a spot in the subsurface where its further upwards percolation through the reservoir is somehow blocked (a petroleum *trap*), will start to accumulate in the fine pore spaces of the reservoir rock layer. Such a trap could be formed at the crest of an anticline.

This is where we must seek petroleum and produce it from. It is not found in great caverns beneath the ground but is subtly controlled by the rocks and their characteristics we have been considering. The challenge to the explorationist is to predict where all these necessary requirements are likely to be met, deep underground where he cannot see the rocks!

2

The occurrence of petroleum

2.1 THE NATURE AND FORMATION OF PETROLEUM

Except in the case of dry gas (see below), petroleum consists of variable mixtures of the chemical substances known as *hydrocarbons*. The molecules of these, as the term suggests, are made up predominantly of atoms of carbon and hydrogen in varying abundances, arranged and linked together in many different ways. Some other elements may also be present: oxygen, nitrogen, sulphur, and metals such as vanadium, copper, and nickel. The molecules can be large and extremely complex. They are commonly described and depicted by semi-graphical chemical formulae, of which a few of the simplest are shown in Fig. 2.1; the formulae of complex ones, even using a form of short-hand, may cover a page!

Many hundreds (literally) of these compounds are present in a single crude oil, and it is their variety that accounts for the wide range in the properties and appearance of oils. The majority of the hydrocarbons fall into three families, the names of which may be familiar to the non-chemist (Fig. 2.1):

1. The *paraffins* or *alkanes*, in which the carbon atoms are arranged in straight chains, some of them branched.

2. The *naphthenes* or *cycloalkanes*, where five or six carbon atoms are arranged in a ring to form the basic structure.

3. The *aromatics*, in which six carbons are also arranged in a ring but bonded together in a different way.

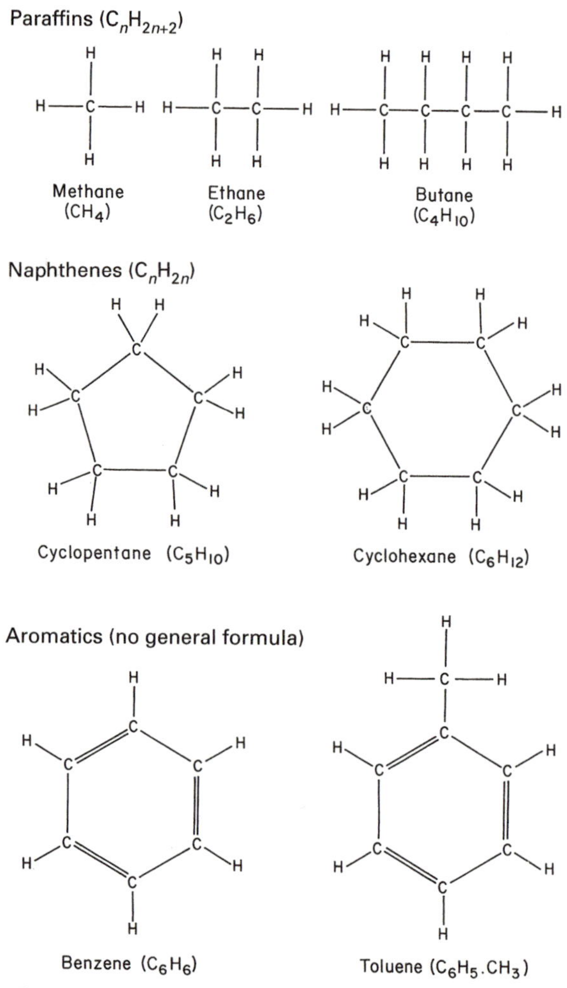

The various components of a naturally occurring crude oil have their different uses, and hence values. Petrol is familiar to us as one of the most valuable, but it may have come from the same crude as diesel oil, lubricating oil, candle wax, and the raw materials of plastics, to mention but a few. It is the function of a *refinery* not only to separate out the various hydrocarbons actually

Paraffins (C_nH_{2n+2})

Naphthenes (C_nH_{2n})

Aromatics (no general formula)

Fig. 2.1 The chemical formulae of simple representatives of the main families of hydrocarbons, showing how carbon and hydrogen atoms may be bonded together. Note that the four simplest of the paraffins (alkanes), containing 1 to 4 carbon atoms, are the only hydrocarbons that occur as gas at atmospheric temperature and pressure; the rest are liquids.

present, but also to break down some of the heavier ones into lighter and more useful products—the process known as *cracking*. Because of the variation in composition, crudes from different parts of the world may have to be treated in separate refineries designed to particular specifications. Detailed analysis of newly discovered oil is thus essential to determine just what it consists of, and where and how it should best be refined; this in turn will establish its potential value.

Of more immediate importance to the explorationist, perhaps, is that an understanding of its composition can give clues as to where, and possibly how, the crude oil was generated and what may have happened to it under the ground subsequently. This in turn can help to direct our exploration to more closely defined areas.

2.1.1 Natural gas

Natural gas occurs in the ground either on its own or together with oil. In the latter case, some of it will be dissolved in the oil under subsurface pressures and temperatures but, if there is enough of it, it may separate out in the reservoir as a *gas cap* above the oil. Gas that is dissolved in the oil underground will bubble out when the pressure is released as it is brought up to the surface, just as carbon dioxide bubbles out of soda water when it is poured into a glass. There is often not enough of this *associated gas* to justify the cost of a separate pipeline to take it to market, so what cannot be used locally or on site must be disposed of safely. Increasingly these days, surplus gas is compressed and pumped back down underground to assist oil production (see Section 4.4), but otherwise it just has to be flared—wasteful but unavoidable!

Gas is often found entirely on its own, with no oil in the vicinity at all; this is the case in the southern North Sea, which supplies most of the gas that is distributed to homes in Britain. In such conditions, the gas is usually 'dry', that is to say it consists almost entirely of methane, the simplest and lightest of the paraffin family of hydrocarbons (one atom of carbon with four of hydrogen, CH_4). Gas associated with oil, on the other hand, is commonly 'wet' with minor and decreasing proportions of ethane (C_2H_6), propane (C_3H_8), and butane (C_4H_{10}). These four hydrocarbons are the only ones that occur as gas at normal surface temperatures and pressures; they can, of course, be liquified under pressure, as for example in cooking gas cylinders.

There are also certain mixtures of light hydrocarbons, which occur as gas under pressure in the Earth but condense to very light oils when produced to surface. These are the so-called *condensates*. They can cause a lot of problems to the production engineer, but they are valuable; some could be put directly into the tanks of motor cars.

2.1.2 Crude oil

Naturally occurring crude oils vary enormously, depending on the types and relative proportions of the hydrocarbons forming the mixture. They range in colour from pale yellow, through a series of greens and browns, to black. They also vary from light, very mobile runny liquids to thick sticky ones that will scarcely flow; in more scientific terms, they vary in both density (or specific gravity) and viscosity.

In Europe, oil densities may be expressed by the specific gravity, but a more commonly used scale is in degrees API (American Petroleum Institute). This is given by the formula:–

$$^\circ API = \frac{141.5}{\text{Specific gravity}} - 131.5$$

The lighter oils have the higher °API values: very light oils and condensates are 50°+, whilst those less than 20° are heavy and, below about 15°, may not be economic to produce. Most oils fall within the 20–40° API range.

Variations in oil quality partly reflect its origin (see next section), but also what has happened to it during the often long period that it has been moving around, or *migrating*, underground. During such migration, discussed in Section 2.2.2, the lighter components of the crude are susceptible to being degraded or lost, leaving a heavier mixture behind. Conversely, deeper burial to hotter levels of the Earth can cause the cracking of some of the heavier components; this is the same process that we have seen used artificially in a refinery. In general, the deeper the oil is found, the lighter it is likely to be; indeed, beyond certain temperature conditions, all of the oil molecules may be cracked and only dry gas (methane) remains.

During its migration, some of the oil often manages to escape up fault planes, or perhaps cracks in the rocks, to the ground surface where its lighter components will evaporate and be lost for ever. Such *seep-*

ages are known from many parts of the world (Figs 2.2 and 2.3), and incidentally they demonstrate that Nature is very efficient in cleaning up her own pollution! A heavy tarry residue may be left, the presence and uses of which have been known since early Old Testament times; the first record is of Noah using pitch to caulk the ark. Gas seepages also occur and again have been used since early times, as poor Shadrach, Meshak, and Abednego could have testified from the 'burning fiery furnace'!

Enough of the heavy residue in seepages may remain to be dug out and used perhaps for road surfacing; the pitch lakes of Trinidad provide a well-known example. Sometimes also, such material is found impregnating sands or sandstones at the surface to form *tar-sands*. Vast amounts of very heavy petroleum exist in the tar-sands of Western Canada, Venezuela, and Malagassy; technically it can be extracted and processed, although its working has not yet proved viable economically. Possibly we have here a resource for the future, when the oil price justifies exploitation.

The importance of seepages to exploration is that they demonstrate without doubt that at least some oil or gas has been generated in the vicinity and that it is able to migrate in the subsurface. This in itself may be a sufficient encouragement for us to continue further exploration. Indeed, many early discoveries were made by drilling close to surface seeps, and some companies still direct considerable effort to the remote detection, by the so-called 'sniffing techniques', of trace amounts of escaped petroleum.

Fig. 2.2 A natural oil seepage near Gisborne, New Zealand. Oil trickles up from below to cover about twice the area shown. It partly evaporates and partly becomes degraded, so that it does not spread further.

Fig. 2.3 A small oil seepage near Weymouth, southern England. The oil impregnates a bed of sandstone in the cliff face, and trickles down the rock face below. Note hammer for scale. Photograph by R.C. Selley.

2.1.3 The generation of petroleum—source rocks

The petroleum that we are exploring for and exploiting was formed by the action of the Earth's heat on the remains of ancient plant and microscopic animal life incorporated and preserved in sedimentary rocks.

Trace amounts, especially of methane, are added to the Earth's outer layers by emanations from deep in the molten interior: methane has been identified among the gasses given out by volcanoes. However, various lines of evidence establish conclusively that our petroleum resources have an organic origin. They include the way in which oil and gas are distributed within sequences of sedimentary rocks, the presence of some petroleum molecules that are barely altered from ones that occur only in living organisms (chlorophyll, for example), the physical properties of some oils relating them to living matter, and the proportions of different carbon isotopes. Do not be misled by the siren voices of the advocates of an unlimited deep Earth source for all of our petroleum!

The kinds and amounts of petroleum that are generated in the Earth are governed by three factors:

1. the nature of the remains of living organisms preserved in the sediments;
2. the abundance of this organic matter; and
3. the extent to which it has been heated (cooked) through burial.

Let us look at each of these three factors.

The nature of the organic matter The living material that we are concerned with does not come from the fishes and other larger animals that are familiar to us; even where they are especially prolific, there are just not enough of them. Of interest to us are the very much more abundant, minute primitive organisms that require a microscope to be seen. This organic matter has been classified into three types (Table 2.1):

Type I is from the very fine algae that flourish in some fresh-water lakes, particularly in warmer climates. Where it is especially abundant, it may be seen as a green scum on the surface of the water. When later buried and heated, it gives rise to high quality but waxy oils.

Type II is found in marine sediments, and consists of the remains of the single-celled plankton, algae, and bacteria that live in abundance in certain favoured regions of the ocean (Fig. 2.4). Voyagers in the tropics will have noticed a phosphorescent wake behind the

Table 2.1 The three types of organic matter that can give rise to petroleum, the environments in which they occur, and their principal petroleum products.

Type	Composition	Initial product	Origin	Environments
I	High hydrogen Low oxygen	Waxy oil	Freshwater algae	Lakes
II	Intermediate hydrogen and oxygen	Low-wax oil	Marine algae, plankton	Seas
III	Low hydrogen High oxygen	Gas Some waxy oil	Land plants	Swamps Nearshore marine

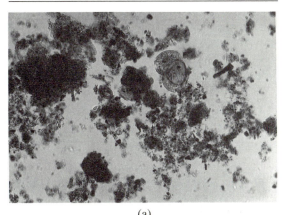

(a)

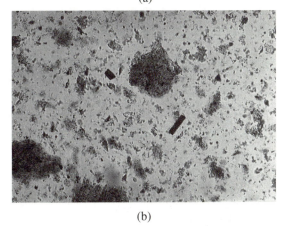

(b)

Fig. 2.4 Two photographs of organic matter extracted from potential source rocks, greatly magnified under the microscope. (a) From Recent sediments in Spain; the mushy looking material is Type II algal (sapropelic) kerogen, the black fragments with sharp outlines are bits of land plants, and a pollen grain is near the top. (b) Type II organic matter from the Jurassic of southern England, together with some land plant fragments. Photographs by M. Rahman.

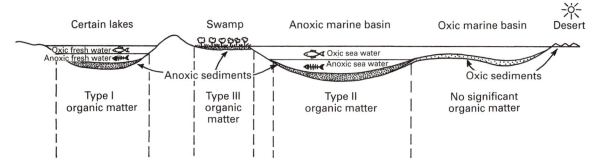

Fig. 2.5 The nature and environments of deposition of the types of organic matter that give rise to petroleum.

ship at night: it is due to a multitude of some such minute organisms giving out a flash of light in annoyance at being disturbed. One environment where this is particularly notable is where there is an upwelling of cold, nutrient-rich deep oceanic waters, as off the coasts of Namibia, California, and Peru: the nutrients support not only the plankton, but in turn the whole life-chain which browses on it, leading to rich fishing grounds. This type of organic matter is the source of the greater part of the world's oil.

Type III consists of material derived from land vegetation, including both wind-blown spores and pollen, and the fragments of plants incorporated into the sediments; coal is the extreme example, consisting of plant remains preserved originally in swamps. This material gives rise primarily to dry gas (methane); the danger of it causing explosions in coal mines is ever present. The cuticle of some thick-leaved plants which flourish in the tropics, rubber and mangrove trees for example, can however produce waxy oils on heating.

Here, then, may be a first guide for exploration. If we can identify the nature of organic matter in a rock, or even infer its nature from the environment of deposition of the sediment, then we may have a clue to the type of petroleum that we may encounter (Fig. 2.5).

Clearly, if this organic matter is going to be converted later into petroleum, it must be protected from destruction as it falls through the sea, at the sea bottom, and during early burial in the sediment. It is liable to be destroyed either by being eaten by animals living in the water or at the sea-floor, or by being oxidized. Preservation from either fate is likely to be ensured if the sea- or lake-bottom is deficient or lacking in life-supporting oxygen (*anoxic*). This can occur in a variety of sediments and environments.

However, it is common only in shales and some very fine-grained limestones deposited under the prescribed conditions, because the coarser sediments may allow later percolation by oxygen-bearing waters.

Potential *source rocks* for oil and gas are thus shales and some fine limestones, deposited and preserved in anoxic environments. Such rocks are usually black, sometimes sulphurous, and lacking the remains of bottom-living creatures; only the fossils of near-surface free-swimming creatures will be found (Fig. 2.6). The original fine layering of the sediment is commonly preserved, not having been burrowed or churned over by

Fig. 2.6 Fossil of a free-swimming ammonite (*Lytoceras* from the Jurassic, related to the present-day octopus) such as might be found in a petroleum source rock: the remains of bottom living creatures are notably absent. Photograph by A.C. Cash.

scavenging animals. Such shales contrast with, for example, grey structureless mudstones containing heavy fossils and having little or no source potential.

Abundance of the organic matter   For a shale actually to act as a source rock, it is generally accepted that the organic matter must amount to at least 0.5 per cent by weight. However, the source rocks of most oil-bearing regions, such as the northern North Sea, have organic contents of at least 5 per cent, sometimes even more.

Occasionally, the figure may exceed 50 per cent, in which case the rock is referred to as an *oil shale*. It may be economic to quarry or mine this and to heat, or retort, it to generate oil artificially for extraction—the oil shales in the vicinity of Edinburgh provided the first serious oil producing activity in Britain during the nineteenth century. Generally, however, it is not economic at the present time to exploit oil shales, and large scale quarrying is anyway often unacceptable environmentally. It is these considerations that have enforced the shelving, for the time being at least, of plans to develop vast oil-shale deposits in Colorado, USA.

Not all of the organic matter that is in the source rock gets converted into petroleum; at best, the figure amounts to some 70 per cent. And not all of the oil that actually is generated ever manages to leave the source rock. The 'expulsion efficiency' will depend on the stage of generation reached (see below), the thickness of the source rocks, and whether or not there are nearby pervious layers to allow the oil to trickle away.

Nevertheless, the organic richness of a possible source rock is the second thing that we shall need to know about it. This may enable us to make some calculations as to the amount of oil that may be present in an area: is it likely to be sufficient to justify further exploration expenditure?

The generation of petroleum   It is only after the source rock has accumulated on the sea or lake floor and started to be buried by a cover of younger sediments, that the changes leading to the production of petroleum start to happen. Let us follow the burial of a typical source rock containing Type II organic matter downwards into the Earth and note what takes place (Fig. 2.7).

The first thing that happens to the organic matter is that it gets attacked and decomposed by bacteria, those 'anaerobic' bacteria that live without oxygen. They break most of it down into a structureless substance known as *kerogen*, which is the actual raw material of petroleum, and at the same time liberate methane. This is familiar to most of us as the 'marsh gas' stirred up with a stick or by our toes from the bottom of a muddy pond; it is known more scientifically as *biogenic gas*. Just occasionally, enough of this is generated and preserved during burial, to accumulate later as commercially exploitable reserves of dry gas: the Cook Inlet region of southern Alaska provides more than enough of it for the city of Anchorage.

As the temperature increases with deeper burial, the bacteria are killed off but then nothing much happens for a while. Once it reaches about 150 °F, however, oil starts to be formed from the kerogen. Initially it is heavy oil molecules that are generated but, as the temperature rises further, progressively lighter oils are produced and the molecules formed earlier are broken down, or cracked, to lighter hydrocarbons. As temperatures continue to rise, we reach a point of peak generation, after which oil production drops off; as it does so, the decreasing size of the molecules formed means that we are starting to get wet gas. Beyond some 350 °F, only dry gas is still generated and pre-existing heavier hydrocarbons are cracked right down to methane: this gas is described as *thermogenic*. Finally, above about 450 °F, all hydrocarbons are destroyed and the rock itself starts to be changed or metamorphosed.

This story means that there is a limited temperature/depth range in the Earth within which oil is generated and preserved; this is sometimes referred to informally as the *oil window* or the *oil kitchen*. A source rock must have been buried this deep to have generated any petroleum at all, and we cannot expect to find oil preserved below the oil window.

It is not strictly correct to relate these changes just to temperature, since time also has an effect. We can get the same results by heating the organic matter to higher temperatures for shorter periods of time (do we put the Sunday joint in the oven at a low setting before we go to church, or at a higher temperature when we get back home?). For this reason, and also because temperature at a given depth varies with location, neither can we relate the process precisely to depth of burial. We therefore refer to the level of *maturity* (the process of cooking is known as *maturation*) reached by the organic matter, and relate this to the stages of petroleum generation. For the scientifically minded, maturity increases exponentially with temperature and linearly with time. To give a general idea, however, the oil window is commonly at depths between some 7000 and 15 000 feet, although these figures may vary even by thousands of feet.

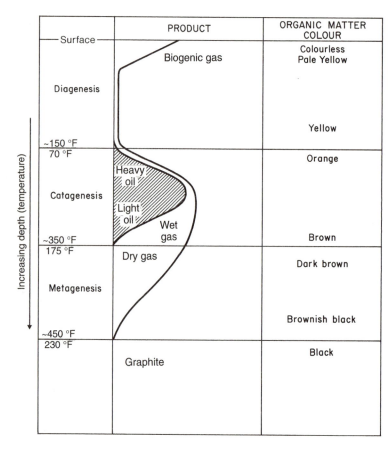

Fig. 2.7 The process of maturation of Type II organic matter and the generation of petroleum with increasing depth of burial and temperature.

Maturity can be measured and monitored by various physical and chemical techniques. A commonly used one measures *vitrinite reflectance*, or the shininess of small pieces of vitrinite, which is the shiny stuff one can see in a lump of coal: the shinier it is, the more mature it is. However, one of the simplest and most effective of the available techniques is just to note the changes in colour of the organic matter during maturation. When fresh it is colourless, becoming yellow (the colour of lager) as it is heated; in the oil window, it is orange to light brown (bitter beer); below this but still producing methane it is dark brown (brown ale); and when it is black (stout) it has been cooked out; as ever, good British bitter is best!

These results, and the study of the maturation reached by possible source rocks, are very widely used in exploration. Has our source rock been cooked enough to generate oil at all? If so, can we expect to find heavy oil, or light oils? Might there be only gas?

These are likely to be questions that are critically important to exploration. If we do not have samples from deep underground, from wells, to make maturity measurements on, then the geologist will have to make predictions about their answers: techniques are available for him/her to do so, and are applied to most exploration programmes. This really is a multidisciplinary exercise, involving at least geology, geophysics, and geochemistry.

2.2   PETROLEUM MIGRATION AND RESERVOIRS

Warning: this section may seem a bit cerebral and dull, and may be somewhat harder to understand than the rest of the book; it is not likely to provide the world's most entertaining reading. But *please* persevere and try not to be put off. It is not just an integral part of the story, but it is absolutely fundamental to the everyday work of many people in the industry. In later chapters, we shall be coming back frequently to what is in it, and some understanding is essential. We will try to keep it as simple as we can.

Oil or gas, when newly generated in the source rock, is too dispersed and thinly spread to be exploitable; in addition, there is no way that, having been generated, it could flow quickly enough through the compact rock into a well for it to be produced. Somehow or other it must be concentrated in an environment from which it can be extracted.

This means that the petroleum must move (*migrate*) from the source rock, where it can be expected to be in the form of very minute droplets, into a rock layer, a *reservoir* formation, through which it can percolate, in which it can be stored, and from which it can be produced. What makes a reservoir are the minute holes, gaps, or *pores* between the grains of, say, a sand or sandstone; it is these that make the rock pervious or *permeable*. Oil does not occur in lakes or caverns under the ground (unfortunately), but in such pores, often too small to be seen with the naked eye, yet large enough to do the job.

We are therefore looking at a two-stage process. Firstly, the oil has to move through and out of the source rock, which is commonly completely tight and impervious: this is known as *primary migration*. Then, once out and into an environment where it can move more freely, it will percolate through the fine channels provided perhaps by the pore spaces of the reservoir—*secondary migration*. Sometimes, as we have already seen, this may take place through cracks or fractures in the rocks such as might be provided by faults. Always though, the petroleum, be it oil or gas, will be down there together with water, which may have been retained from the original sea or lake in which the sediments were deposited, or which may have infiltrated the rock subsequently. Petroleum, being lighter than water (usually), tries to float upwards through it towards the Earth's surface.

These, then, are the constraints under which we have to try to understand the problems of migration.

2.2.1 Primary migration

Understanding just how oil or gas moves through the extremely fine-grained and tight source rock is one of the outstanding problems of petroleum geology. We cannot get down there under the ground to see it happening, and we cannot simulate it confidently in the laboratory because of the long periods of geological time that it may take. People have asked: does it matter? These days it does; we would dearly like to know how it happens, to help us in understanding when and where the oil may have moved, and particularly in trying to estimate the quantities that may be involved. However, we have to admit that we don't really know and that we can only resort to theorizing.

The problem is that, by the time a source rock is buried to the depths of petroleum generation, it has been compacted by the weight of the overlying sediments into a pretty firm and compact rock. We cannot normally get a fluid to enter or move through it at all. And yet it is clear that, somehow or other, the oil just must have migrated because that is where it was generated.

When a mud or clay is first deposited, it may contain as much as 80 per cent water. The process of compaction represents the squeezing out of this water, so that a measure of the water content of a shale indicates the degree of compaction that has taken place. In this way, we know that the greater part of the compaction takes place soon after burial and that the decrease in water content with depth is exponential. By the time the rock reaches depths where oil is generated, there is very little of the water left. This is a pity, because it would be nice to be able to think of the oil being carried away or flushed out with the water. This could have happened in the case of the early formed biogenic gas, which might even be dissolved in the water. But, in the case of oil or thermogenic gas, we are stuck with a problem.

These topics are mentioned, not to confuse the non-technical reader, but in order to describe a phenomenon that we should all be aware of.

It can happen, particularly in thick sequences of shales, that even the water does not manage to escape from the rocks as they are buried. There will be more of it held there than there should be for that particular depth. It will tend to hold apart the small grains of the rock, and the water itself starts to carry some of the weight of the overlying sediments; it therefore builds up higher pressures than are normal for that depth. Clays in this state are said to be *overpressured* and *undercompacted*. Unless they are expected when drilling, and suitable precautions taken, they can be extremely dangerous. There have been instances in many parts of the world, including the North Sea, when unexpectedly high pressures have started to force the drilling mud (see Section 3.3.1) back out of the well; if not prevented, even the heavy drill-pipe can be blown out (Fig. 2.8). This could lead to the sloppy, undercompacted clay flowing all over the drilling rig. Worse, if there is any gas mixed up with the formation water, this could catch fire at the surface and cause considerable destruction and loss of life. Such *blow-outs* used to be a common hazard; however, with today's technology, fortunately they are rare, but they still can happen. It is essential to try to predict the presence of overpressures in advance of drilling and encountering them, so that proper precautions can be taken in time.

There are various ways in which overpressures can be forecast, but a useful clue is provided in some land areas by the occurrence of *mud volcanoes*. If there is a line of weakness, such as a fault, in the sediments above the overpressured shales, then these runny muds

Fig. 2.8 A spectacular blow-out of high pressure gas in Iran in 1951. The escaping gas caused a flame 1200 feet high and blew the heavy steel drill-pipe out of the well and into the air. Such accidents are to be avoided if at all possible! Photograph by British Petroleum.

may burst through and up to the surface. There they dry out and build up little mounds with a small crater in the middle, for all the world like a mini volcano (although nothing whatsoever to do with real volcanoes). There are beautiful examples in New Zealand (Fig. 2.9), Trinidad, and south-east Asia, to name but a few.

To come back to the question of primary oil migration, it is possible that overpressuring may help us to understand the problem. Any shale may, to some extent, become overpressured and relatively undercompacted during its burial; if this is still the case when the oil is generated in a source rock, then the oil might not have such a hard time in moving through it. But anyway ... we are back to speculating!

2.2.2   Secondary migration

Once the petroleum has made its way out of the source rock, we can imagine it being much freer to migrate, as small droplets, large drops, more coherent slugs, or even as continuous 'streams'. The way in which it now moves is much more understandable and predictable. Basically it tries to move upwards through the water which otherwise fills the permeable channels, just as a drop of oil will float up towards the surface of a container of water. It is lighter than the water and hence buoyant in it.

What oppose this movement are the strong forces acting at the narrow gaps between the pores of the rock—the capillary forces. An analogy (not a very scientific one!) is to image a crowd of people trying to push through a narrow gate; the movement of the crowd as a whole is obstructed and it may get jammed up altogether, until the bigger push of the growing crowd forces the few in front through. Similarly, if the buoyancy of the oil drop is less than the capillary forces at the pore throats, then it will stop migrating; but if more oil is added to it, then it may develop enough buoyancy again to move on through.

In this way, oil will work its way gradually up to the surface, where it is lost in a seepage (see above), if it can. If, however, our permeable reservoir layer is overlain by an impermeable bed, known as a *cap rock* or *seal*, then obviously the oil will be stopped. And if, as it commonly is, the reservoir is sloping or dipping, then the oil may be able to move further upwards by migrating laterally sideways at the same time. If its path up towards the surface is completely blocked, by a *trap* (see Section 2.3), then the oil has no choice but to stay there until more and more drops arrive to form

Fig. 2.9 Small mud volcanoes near Gisborne, New Zealand. Liquid, overpressured mud reaches the surface from considerable depth, and dries out to form the little volcano-like mounds.

an accumulation or *pool*. This, of course, is where we have to look for it.

With most accumulations, there is evidence that the petroleum has not migrated laterally for more than a few miles. In some fields in western Canada, parts of South America, and elsewhere, however, it is clear that we are looking at oil that has migrated well over 100 miles, some of it in the case of Canada from actually underneath the Rocky Mountains to the west. This again is something that has to be considered when we are trying to find the elusive 'black gold' in the subsurface.

There is yet another possible difficulty. Water can flow through a permeable reservoir just as well as, or better than, oil. The fountains in Trafalgar Square in London used to be fed entirely from wells drilled down into the pervious Chalk, to tap the water that had entered it as rain in the Chiltern Hills some 20 miles away. The fountains kept playing as long as there was a good flow through the Chalk. Nowadays we are using so much of this water for other purposes that there is sadly no longer enough to feed the fountains which are maintained by pumps. The flow of water through a reservoir that has oil in it is sometimes strong enough to wash or flush the oil out altogether and, for this reason, a promising prospect may turn out to be unsuccessful when we drill it.

If the reader has got this far, he/she will no doubt be thinking that all this is very complicated and a bit esoteric: he/she will be in good company. It is very difficult to understand just how and where oil might be moving around down there underground where we cannot see it,

let alone to predict where it may have gone to. That ideally is what the petroleum geologist should be able to do; but, in practice, we cannot do much until we have a lot of information and knowledge about an area, and one has to admit that this is a part of our predicting that often tends to be taken for granted or even ignored. Don't blame the geologist: he does what he can.

2.2.3 The reservoir

Enough has been said already to point out that petroleum occurs in the subsurface in the small holes in porous rocks. These holes are usually the minute pore spaces between the grains of the sediment, although sometimes they are found to have been dissolved (leached) out after burial; just occasionally, the space is provided by cracks and fractures in otherwise impervious rock. The most common reservoir beds or formations are provided by sandstones and carbonates.

We must, however, look more deeply into the nature and properties of reservoirs and their significance.

To be effective, a reservoir must have three essential features.

1. It must have holes in it to contain the oil or gas: the greater the pore space (*porosity*), the larger the amount of petroleum present will be. But note that we are only interested in holes that are linked together, i.e. in the *interconnected porosity*. Fluids could get neither into nor out of holes that are isolated.

2. The pore spaces must be sufficiently interlinked to allow the petroleum to move through the rock and into the well. The higher the *permeability*, the higher will be the rate at which the well can be produced: if it is too low, then it may not be economic to use the well for production.

3. The reservoir formation must be sufficiently thick for there to be enough oil or gas to be worth producing.

Porosity (usually designated by the Greek letter ϕ) is defined simply as the volume of the pore space in proportion to the total volume of a sample, expressed as a percentage:

$$\phi = \frac{\text{Volume of pore space}}{\text{Bulk volume}} \times 100 \text{ per cent}$$

Strictly speaking, we are concerned with the interconnected pore space, which defines the *effective porosity*.

To see how this works, take a jar filled with dry ordinary sand and a similar jar filled with water to the same level as the sand. Gradually pour the water into the sand until it is completely saturated and free water starts to appear on top. How much water has gone into the sand? Probably nearly half of it, suggesting that about half of the space occupied by the dry sand was empty: the porosity was nearly 50 per cent. Depending on the type of sand used the porosity may be considerably less than this, or even zero if there is a lot of mud mixed up with it. In a sandstone reservoir, the principle is exactly the same, except that the grains may be lightly cemented together; the space that still remains is the porosity of the rock. This story is developed in the next section.

Unfortunately, in most reservoirs the picture is a little more complex than this. We commonly find that there is a very thin film of water sticking to the surfaces of the grains (Fig. 2.10). This water was there before the oil arrived, and may be a remnant of the water under which the sediment was originally deposited; it is held there by very strong interfacial tension and atomic bonding, and it cannot be shifted. This water regrettably occupies some of the pore space that we would prefer to see filled with oil, and we must allow for it in, say, our reserves calculations. The percentage of the porosity occupied by this immovable water is referred to as the *water saturation* or the *irreducible water saturation* (Sw). The space that can be filled by petroleum, then, is given by:

Fig. 2.10 Sketch of sand grains greatly magnified, showing the way that immovable water films cling to them. This irreducible water saturation reduces the porosity available to be filled by petroleum; its percentage will be governed by the grain-size.

$$\phi \times (1 - \text{Sw})$$

These quantities, porosity and water saturation, are so important to us that specialists, or teams of specialists, using complex computer programmes, may devote considerable time and effort (even their careers) to determining them, from well logs (see Section 3.3.2) or from direct measurements in the laboratory on suitably prepared samples taken from drilled wells. Petroleum engineers especially will base a lot of their work on the results obtained.

Permeability (*K*) is equally important. It is a measure of the ease with which fluids can move through a rock: force them through and see how fast they come out of the other end of a sample. This is, in fact, how permeability is defined and measured: by doing such an experiment in the laboratory on a specially cut cylinder of the rock. The permeability, in *darcies* (or more usually millidarcies), is calculated from the formula which defines it:

$$Q = K \cdot \frac{Ap}{\mu l}$$

where Q is the quantity of the fluid that comes through in unit time, A is the cross-sectional area of the sample of rock, μ is the viscosity of the fluid used, p is the pressure drop between the ends of the sample, and l is the length of the sample.

It is not really as complicated as it might seem. But it is interesting that K, the permeability, turns out (or ideally should do) to be the same whatever fluid we use to measure it; it is a property of the rock, the viscosity of the fluid offsetting the rate of flow through our sample (μ and Q in the formula).

What does affect the permeability, however, is the presence of another fluid in the rock. We noted above that there is usually water in there, the irreducible water saturation, and this can have a dire effect on the permeability as far as any oil is concerned: the greater the water saturation, the less permeable the reservoir will be to oil or gas. This problem need not worry us here, but it is a matter of considerable concern to the petroleum engineers working out, for example, the rates at which wells can be produced; if one hears them talking about *effective permeability* or *relative permeability*, this is what they are wrestling with.

The *thickness* of a reservoir rock layer, or *formation*, must be considered when we are trying to decide whether or not it is going to yield enough oil or gas to justify our well and production costs. Precisely what we regard as 'thick enough' will depend on many factors: How porous is it?—in other words how much oil could it contain? How permeable is it or at what rate will the oil flow out into our well? Are there other reservoir beds nearby in the succession that could also contribute to production? Where in the world is the well?—if it is in the yard of the refinery, a relatively thin reservoir may be sufficient but, if the oil has to be brought from beneath the sea or from a tropical rainforest, we shall need a lot of it to cover the costs.

What we are saying is that there can be no simple answer to how thick a reservoir we need. However, we shall certainly have to make estimates of how much oil there is down there in the ground (see Section 3.5), and this inevitably will involve knowing how thick the reservoir is, as well as its porosity and permeability.

The majority of the reservoirs that produce oil and gas are provided by sandstones and carbonates, and it is necessary to know a little more about them.

2.2.4 Sandstone reservoirs

Unfortunately, not all sandstones form satisfactory reservoirs. They need to have certain characteristics if they are to contain and yield sufficient petroleum to be useful. These are characteristics that we may need a microscope to see, but perhaps we can appreciate what is going on by picturing the rock in our minds magnified so that the grains look like tennis balls.

If the grains of the rock are poorly sorted according to size, then the smaller ones will tend to occupy the spaces between the larger ones reducing the porosity. Similarly, if there is some clay mixed in with the sand grains, this will block up the pore spaces and the gaps, or throats, between them. The same sort of effect can be produced by cement precipitated after the sand was deposited, from calcium carbonate- or silica-saturated water percolating through the rock (remember the scale inside a kettle, or blocking up hot-water pipes). In addition, it turns out that porosity is highest if the grains are well rounded, so that sharp corners on them do not jut out into the pores: think of a pile of tennis balls and the gaps in between them.

So the best reservoir sands are those in which the grains are all of the same size; are well rounded; devoid of clay and only lightly cemented so that the grains are just held together; and preferably coarse so that those immovable water films sticking to them do not take up too much of the available pore space. These conditions are often best met by sands that were deposited as dunes in a desert (Figs 2.11 and 2.12): grains blown around by storm winds tend to have their corners smoothed off, and they are nicely size-graded by varying strengths of the wind. Much of the gas in the southern North Sea is contained in the subsurface in such dune sands of Permian age (about 250 million years old) (see Section 5.3.1).

In a cliff, a sandstone may appear to be all much of a muchness. However, most sands do in fact vary quite a lot both in vertical sequence and in lateral extent: variations in texture and composition can often be seen quite clearly on close inspection. Take a look, for example, at the photographs in Chapter 1 and at Fig. 2.13. Such variations are all important; they have a strong influence on the way that fluids will move through the rock, how much oil the reservoir can contain, and how much of it can be produced (see Section 4.3).

Think again about the sand on a beach. It will not extend far inland, if at all, and it may well be that mud is being deposited once we get below the depth of wave action offshore. Sands, and ancient sandstones, are thus liable to be restricted in their lateral extent. These natural variations pose a further problem for petroleum geologists: they not only have to predict that the sandstone is there at all, but they also have to try to forecast its distribution and its variations in quality. This is done by trying to understand the environments that the sand was deposited in, and we mentioned some of the clues that may be used in Section 1.2.1. It may all be deceptively easy if we have only a little informa-

Fig. 2.11 Cross-bedding formed by desert sand dunes of Permian age at Dawlish, Devon, southern England. The general dip is some 15° to the right. As a result of climate change due to continental drift, a vast desert covered much of northern Europe in Permian and Triassic times. Such sandstones form excellent reservoirs, for example for gas in the southern North Sea.

Fig. 2.12 Spectacular desert dune sands of Jurassic age at Zion National Park, Utah. Photograph by A.T. Pink.

tion, provided by a single well for example, but quite often we find that the more control data we get the more complicated an apparently simple picture becomes (Fig. 2.14). Life cannot be described as dull!

That is still not the end of it. We commonly find that changes take place in the rock as it is buried into the deep subsurface. For a start, it gets more and more squashed by the weight of overlying sediments so that the grains may become more tightly bedded down together. Eventually, the weight of the overburden may exert such a pressure, concentrated at the points of contact between the grains, that some of the silica may start to be dissolved and be redistributed into the pore spaces. Porosity is reduced from its original value and, in fact, tails off regularly or linearly with depth (Fig. 2.15). This may affect exploration by imposing a limit to the depths at which we can expect to encounter satisfactory reservoirs. All need not be lost, however.

to effective exploration and to prudent oilfield management.

2.2.5 Carbonate reservoirs

Limestones commonly have high porosity when they first accumulate. There may be a lot of empty space in a pile of shell debris, or in the skeleton of a coral colony. This, however, is seldom preserved for long for, as we saw in Chapter 1, calcium carbonate is particularly prone to dissolution or reprecipitation, depending on the acidity or alkalinity of percolating waters. The original pores are commonly quickly cemented up as the components are welded together into a solid rock. Where we do encounter holes in the rock, they have frequently been dissolved out, or leached, secondarily (Fig. 2.17). Indeed, the rock may go through repeated cycles of leaching and cementation, so that the origin of the pore spaces that we end up with may become pretty obscure. This makes porosity prediction even more difficult than in sandstones, so that in practice we may just have to keep our fingers crossed and hope that it has developed where we want it to be. Sometimes, however, we may be lucky enough to locate a preserved coral reef or pile of shell debris, which may make an attractive target for exploration drilling.

Really good news is where the calcium carbonate has been replaced by calcium magnesium carbonate—dolomite (see Section 1.2.1). This is because dolomite is denser than limestone (it takes up some 13 per cent less space), so the conversion results in the creation of often quite large holes in the rock, which may even be visible to the naked eye. Provided these holes are linked together, the dolomite can provide an excellent reservoir.

Although porosity in carbonates may be difficult to predict, and not all limestones develop even secondary porosity, there are two factors that may make them highly desirable as reservoirs. First, the secondary leaching may create wide permeability channels in limestones as well as in dolomites, which can provide a considerable boost to production. Secondly, a cemented limestone is a rigid and somewhat brittle rock, so that it is particularly susceptible to fracturing which can provide open, even gaping, passageways for oil. The fractured limestone reservoirs of south-west Iran are particularly renowned for their high well production rates.

So, as compared with sandstones, carbonate reservoirs generally tend to contain relatively less oil but to yield it at higher rates for shorter periods of time.

Fig. 2.13 Triassic desert sandstone at Sidmouth, southern England. This rock is cross-bedded near the base, and higher up shows where a channel was cut in the sand and then filled. These features suggest deposition from flowing water during flash floods, producing a reservoir of mixed quality.

Some of the changes in the rock (the technical term is *diagenesis*) may even be helpful since the minerals of certain sand grains can be dissolved away altogether at subsurface temperatures and pressures, and this may create new small holes in the rock (Fig. 2.16). In fact, there is a whole series of such changes that can take place, some of them harmful, some of them helpful to the reservoir.

It is not at all easy to map out and predict these various effects, and we may not have enough information to do so with any confidence until we are well into the development of an oilfield. Nevertheless, an understanding of what a potential reservoir is really like can make all the difference between a successful well and wasting our money by drilling uselessly. Almost certainly, some of our colleagues in an exploration or production department will be concerned with such problems; we cannot overstress their importance

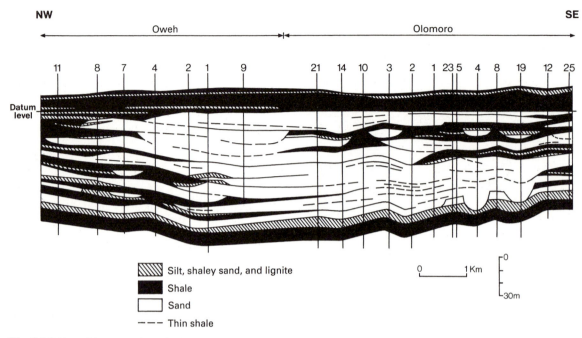

Fig. 2.14 Part of the reservoir section interpreted in the subsurface from a field in Nigeria. The distribution of reservoir quality sandstone is extremely variable, reflecting deposition in and between the river channels of the Niger Delta. After K.J. Weber.

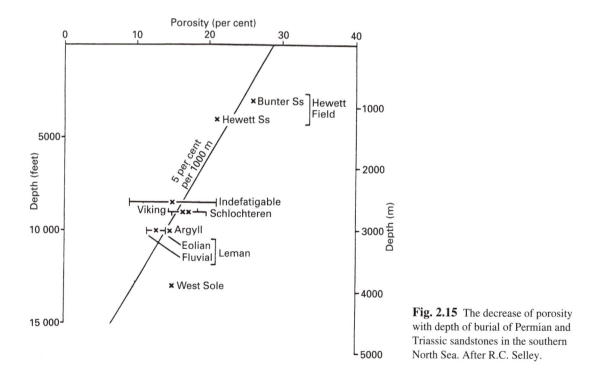

Fig. 2.15 The decrease of porosity with depth of burial of Permian and Triassic sandstones in the southern North Sea. After R.C. Selley.

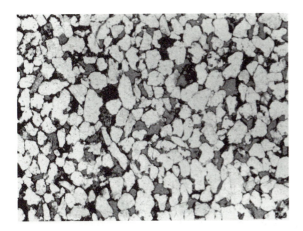

Fig. 2.16 Microscope photograph of a Jurassic reservoir sandstone from the northern North Sea. The grains are white, the cement grey, and pore spaces black. Note that some of the pores appear to have been formed by grains of a different composition being dissolved away. Photograph by R.C. Selley.

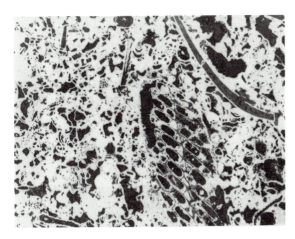

Fig. 2.17 Microscope photograph of a limestone reservoir. All black areas are pores formed where shelly material has been dissolved away secondarily. Note a shell section at top right, a coral-like fragment right of centre, and numerous smaller organisms. Photograph by R.C. Selley.

Their management involves rather different sets of problems and, indeed, may require the expertise of different specialists.

If you have managed to stay with us through this chapter, you should have a good idea of some of the tasks that geologists and engineers are trying to tackle.

2.3 PETROLEUM TRAPS

We must now switch our minds from the microscopic to a very much larger scale. We have seen petroleum generated in and expelled from the source rock formation into an overlying or underlying reservoir. It now gradually percolates upwards through the water occupying the minute pore spaces of the reservoir. If it can, it will escape to surface as a seepage, where it is lost. If then we are to find any of it still preserved, not only must the reservoir be overlain by an impervious layer forming a *cap rock* or *seal* (shales or evaporites are likely to be the most effective), but there must also be some sort of blockage to prevent further migration. This may be caused either by the reservoir itself dying out or by an interruption of its upwards continuity to the surface. Such a configuration of the reservoir is known as a *trap*. Any oil getting there will be unable to migrate further and so it starts to accumulate, by displacing the water already there in the porosity.

The location of a trap in the subsurface is often the first objective of an exploration programme. Indeed, before we reached our modern understanding of the geology of petroleum, exploration used to consist largely of finding a trap, drilling a well into it, and hoping for the best. Nowadays we can do better, and furthermore we can map out the extent and shape of the trap with a good deal of precision—thanks mostly to modern seismic techniques (Section 3.2.3).

2.3.1 The representation of traps

Traps are commonly depicted in two ways. First, they can be mapped by means of contours drawn on the top of the reservoir formation. A *structure contour map* resembles an ordinary topographic contour map, except that now the contours are in depth *below* sea-level, so that the highest points on the map have the *lowest* values. Faults will be marked by jumps of the contours, as the beds on one side are dropped down relative to the other. Examples are shown in Fig. 2.18.

To complement the structure contour map, one or more *cross-sections* may be drawn; these can be regarded as imaginary vertical slices through the Earth and looked at sideways, as in Fig. 2.22. To give a true representation, they should properly be drawn with the same scale for both the vertical and the horizontal, but it is often convenient to exaggerate the vertical to show the individual beds more clearly. Note that we commonly highlight petroleum accumulations by

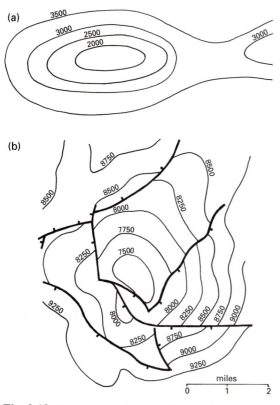

Fig. 2.18 Structure contour maps. The top of a reservoir formation, say, is mapped by contours showing depth below sea-level. (a) A simple hypothetical anticline. (b) A representation of the Piper field in the North Sea (after J.J. Williams, D.C. Connor, and K.E. Peterson): the heavy lines are faults cutting the top of the reservoir and causing the contours to jump; the ticks are on the downthrown sides of the faults. The contours are in feet below mean sea-level.

shading or colouring the reservoir formations where they contain oil or gas, which may give a misleading impression of 'lakes' of petroleum under the ground!

Before we go further, we need a few definitions. These are illustrated on Fig. 2.19 using a simple anticline as an example. The highest point of the reservoir, up towards the ground surface, is known as the *crest* of the trap. The lowest point, which may refer either to its depth or to the spot under the ground where it lies, is the *spill-point*: this is where oil, if more continues to migrate up into the trap than can be accommodated, will spill out (under) and migrate on. The vertical height between the spill-point and the crest is referred to as the *closure*, and the same term is used loosely to refer to the area of the trap above the level of the spill-point.

Oil being lighter than water, separates out on top within the pore-spaces of the reservoir, so that we can recognize a generally horizontal *oil–water contact*. Similarly gas, being lighter still, will occur as a *gas cap* above a *gas–oil contact*. If there is no oil, then we may see a *gas–water contact*.

A single accumulation of oil or gas is called a *pool*. Where there is more than one such pool in the same or overlapping areas, perhaps if more than one reservoir is present, they are embraced by the familiar terms *oilfield* or *gasfield*.

Just a couple more terms. The vertical height of the oil (or gas) between the crest of the trap and the water contact is the *oil-* (or *gas-*) *column* (Fig. 2.20). When referring to a single well, the informal term *pay* is often used. Let us remember, however, that most reservoir formations include some intervals which are *tight*, i.e. which have porosities and permeabilities too low for them to contribute oil to production. These have to be discounted and the bits that remain as useful reservoir in a well section may be lumped together as the *net reservoir* with a *net pay*.

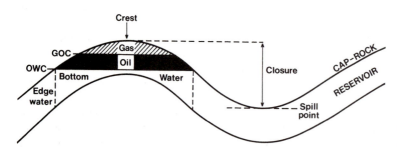

Fig. 2.19 Some terms used to define a trap, using a cross-section of a simple anticline as example.

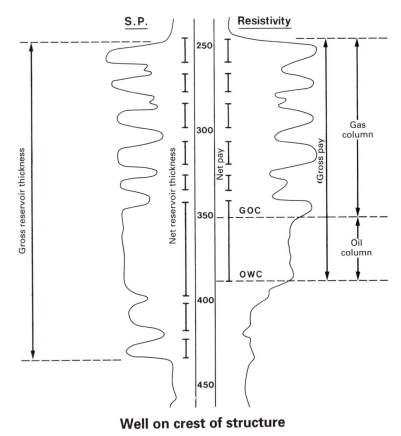

Well on crest of structure

Fig. 2.20 Definitions of net reservoir, column, and pay. The wiggly lines represent wireline logs (see Section 3.3.2); where they bulge out from each other, we are looking at sandstone; where they come together we are in shales. Note also that the oil–water contact may show as a decrease in electrical resistivity.

Now we can start to consider the types of trap whose discovery may await us. They are normally classified under four headings, the first two of which are illustrated in Fig. 2.21:

1. Structural, where the trap has been produced by deformation of the beds after they were deposited, either by folding or faulting.

2. Stratigraphic, in which the trap is formed by changes in the nature of the rocks themselves, or in their layering, the only structural effect being a tilt to allow the oil to migrate through the reservoir.

3. Combination traps, formed partly by structural and partly by stratigraphic effects, but not entirely due to either.

4. Hydrodynamic traps, which are rare and are mentioned mainly for completeness. The trap is due to water flowing gently through the reservoir and holding the oil somewhere where it would not otherwise be trapped.

2.3.2 Structural traps

The best known type of trap is the anticline: on reaching the crest, petroleum migrating up along a reservoir can go no further and it accumulates there as a pool. However, there are various types of anticlines with different shapes and geometries that can affect both their prospectivity and the positions of optimum drilling locations: we have to try to understand them. Traps can also be formed against faults if a chopped-off reservoir is thrown against a shale or other impervious rock.

The general principles of this are straightforward and may be all that the non-technically minded reader really needs to know. But, since our purpose is to portray the work of the explorationist and his challenges, we will describe in a little detail the most important types of anticline, noting the differences in shape and prospectivity that we have to try to interpret. We shall then be briefer on the other types of trap.

STRUCTURAL TRAPS

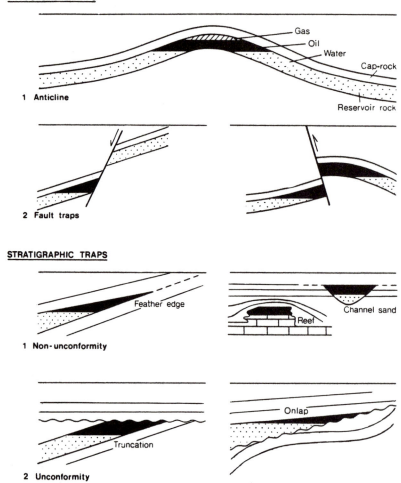

Fig. 2.21 The main categories of structural and stratigraphic traps.

Anticlines Anticlines (and synclines) are most commonly thought of as resulting from lateral compression in the Earth's crust, as described in Section 1.2.3. These *compressive* structures pose one problem right from the start. If, in cross-section, the anticline is asymmetrical, with one flank steeper than the other, then the position of the crest will shift with increasing depth (Fig. 2.22); therefore in order to drill into a reservoir near its highest point (where we would expect the oil to be), we have to know its depth to know where best to locate the well. Seismic may help, but we commonly have to undertake some form of geometrical construction to interpret what is happening at depth. This leads us into the next problem.

Compressive structures have a range of shapes between the purely *concentric* or *parallel* anticline and the *similar* fold, depending on the nature and strength of the rock layers being folded. Let us see what the implications are for exploration.

In the concentric fold the tops and bottoms of all the layers remain strictly parallel to each other, so that the beds maintain a constant thickness throughout. From Fig. 2.23 you will see that these conditions mean that the anticline becomes smaller and tighter at deeper levels until we reach a common 'centre of curvature'. Below this point we have just too much rock to fit into the anticline, so that the beds become intensely crushed and thrust together: we may no longer even have an anticline at all. This means that, in this type of

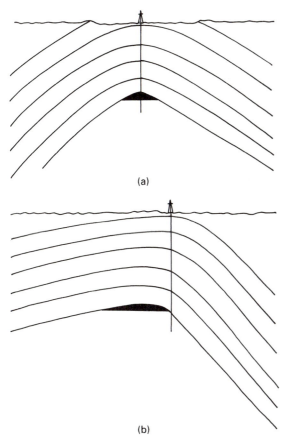

(a)

(b)

Fig. 2.22 Cross-sections of trap-forming anticlines. (a) The dips are the same on both flanks and the crest is beneath the same locality at all depths. (b) The anticline is asymmetrical and the crest shifts with increasing depth. To test the crest at depth, a well would have to be located off-crest at surface.

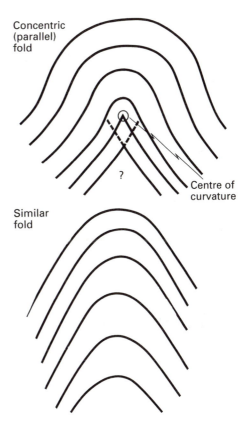

Fig. 2.23 Concentric and similar anticlines. Please see text for explanation.

structure, we can expect to find only smaller and smaller accumulations of petroleum down to the centre of curvature, beyond which there may be no trap left to explore (Fig. 2.24). There is a definite limit to the depths to which we should drill.

The similar anticline, on the other hand, maintains its shape constant down to depth. This can only happen if there is an apparent thickening of some beds over the crest (or axis) of the fold. In this case, we can find the trap present at all levels down to the basement, and we may be able to continue exploration down to depths where we have to stop for other reasons. This is a very different kettle of fish from the concentric anti-cline. In practice, many structures have forms in

between the two extremes, but an understanding of the shape and size of a prospect is clearly critical to programming an exploration well.

Other types of anticline can be formed without any lateral compression at all: an important one is the *drape* or *drape–compaction* structure (Fig. 2.25). Imagine an old-fashioned stone hot-water bottle in a bed with a blanket over it: we can still see the form of the hot-water bottle, and the blanket bulges upwards with an anticlinal shape. Cover it with a few more blankets and a duvet or two, and we may no longer be able to see where the bottle is. Similarly, if the first sediments in a basin were deposited over a hilly surface, or over an upfaulted block or *horst* (the converse, a downfaulted block or trough, is known by another German term—a *graben*), then they will blanket the hill as an anticline; higher beds will gradually mute and suppress the structure until it is no longer present at shallow levels. A second effect

Fig. 2.24 A concentric anticline in south-west Iran. The beds seen on the flanks have been eroded away over the crest, so that deeper formations are exposed. Note that the lower, older beds in the core are compressed into a narrower and tighter fold than the higher ones, and that the anticline would not continue to offer prospects much below the valley level. Photograph by British Petroleum.

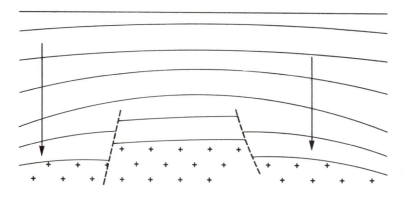

Fig. 2.25 A drape-compaction anticline, the beds being draped over an upfaulted block (horst) of basement rocks. Note that the anticline dies out upwards towards the surface.

comes into play here: because there is a greater thickness of beds off the structure than over the top, those near the bottom of the sequence are going to be squeezed and compacted more on the flanks than on top of the feature as it gets buried. This compaction enhances the anticline formed by the drape; it is not always easy to separate out the two effects, and hence the combined name. In case anyone should think that this is unimportant, note that the largest oilfield in the world, Ghawar in Saudi Arabia, which contains more than four times as much oil as the whole of the North Sea put together, is in one such trap. Another is the

Forties field in the North Sea, where the beds are draped over the eroded stumps of an old Jurassic volcano.

Now we really stretch your credulity and your ability to marvel at the ways of Nature! We pointed out in Section 1.2.1 that a thick layer of salt can accumulate on the floor of a shallow restricted sea under conditions of intense evaporation. Salt under gentle pressure is plastic: it will gradually flow. And salt is also lighter in weight than all other rocks. If, then, we have a layer of salt a few hundred feet thick buried under a cover of a few thousand feet of other sediments, it can start to flow; once such flow is triggered off, the salt being light will try to escape upwards towards the surface. The effect will be initially to bulge up the overlying sediments as an anticline, a *salt dome*, and then to burst through them as a pillar of salt cutting those sediments, a *salt plug*; it may extend up to the surface of the ground or only part way if the supply of salt is limited. Salt plugs are usually more or less circular in plan view, but they can be elongated to form a wall of salt cutting the overburden.

A wide variety of traps can be associated with salt plugs (Fig. 2.26). Not only may an anticline be pushed up over the plug, it is also liable to fracture the overlying and surrounding beds creating fault traps (see below); it may bend up and seal off the strata it cuts through, and finally a residual bulge may be left between two nearby plugs: a *turtle* or *turtle-back* structure. All of these possible traps may contain hydrocarbons. Improbable as the whole story may sound, extensive salt deposits and plugs occur in many parts of the world: the southern North Sea and northern Germany; the Gulf Coast of the USA; the Canadian Arctic Islands; much of the west coast and continental shelf of Africa; the Middle East; and several others.

The last type of anticline that we should be aware of is the *roll-over anticline*. This occurs alongside a normal fault, representing a stretching of the Earth's crust as described in Section 1.2.3. If the fault is curved, so that it is steep near the surface and flattens with depth, then the stretching in effect means that the downthrown side is being pulled away from the upthrown side. This in turn would tend to create an open fissure along the fault. Nature, however, does not like empty holes, and the beds on the downthrown side above the curving fault collapse to fill the gap, bending downwards into the hole. This creates a dip, or roll-over, in towards the fault; a general dip in the opposite direction means that we have thus created the two flanks of an anticline (Fig. 2.27). Note a characteristic of these anticlines: not only do they 'grow' with depth, but they are asymmetrical; at deeper levels the crest will shift away from the position of the fault at surface. Again, therefore, we have to know whereabouts in the succession our prospective reservoir lies, and its depth, to locate an exploration well in the right place.

These roll-over structures are particularly important where the 'stretching' is caused by a very thick pile of sediments at the edge of a continent gently slipping, or

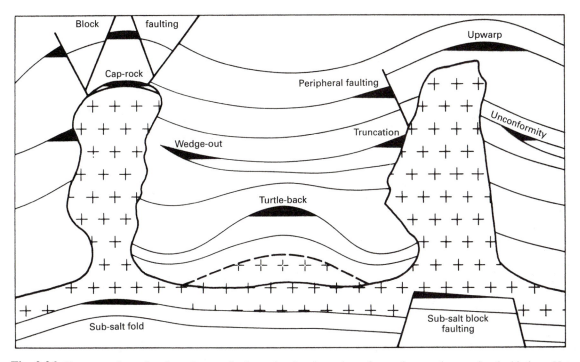

Fig. 2.26 Diagrammatic section through two salt plugs, showing the variety of traps that may be associated with them. Note also that salt, being plastic, can be a perfect seal to any underlying accumulations.

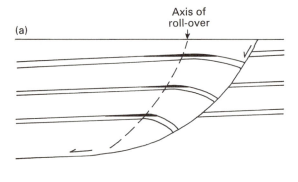

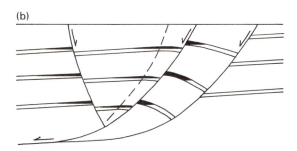

Fig. 2.27 Roll-over anticlines: (a) a simple roll-over into a normal fault; (b) a roll-over complicated by subsidiary faulting near the crest. Note that, in both cases, the position of the crest is displaced with depth and that accumulations in successive reservoirs will not underlie the same surface position.

slumping as a sort of land-slide, down towards the deep ocean. Much of the oil under the Niger and Mississippi Deltas is in such anticlines.

Fault traps We indicated above that a trap may be formed where a dipping reservoir is cut off up-dip by a fault, setting it against something impermeable. The proviso is that we also have lateral closure, commonly at right angles to our line of cross-section: this may be provided by further faulting, or by opposing dips. The large Wytch Farm oilfield of southern England offers a splendid example (Fig. 2.28).

We do not propose to discuss fault traps in detail, although there are many problems in trying to locate them in the subsurface, and in understanding them; Fig. 2.29 may give some idea of what we are up against. Whether or not there is a trap, and how big it is, will depend on the dip of the reservoir as compared with that of the fault, whether the fault is normal or reverse; and it will depend on the amount of displacement on the fault, whether or not the reservoir is completely or only partially offset. The intrigued reader may care to think through the various situations sketched as bits of cross-sections in Fig. 2.29.

Another major difficulty confronts us here. We know that sometimes, as at Wytch Farm, a fault can provide a seal, but we also know that sometimes faults are pathways for migrating petroleum. Occasionally indeed, it seems that one and the same fault may act,

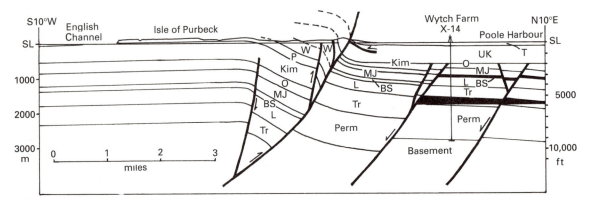

Fig. 2.28 Cross-section through the Wytch Farm oilfield, southern England. The oil is in two reservoirs, trapped against faults to the south; these predated the deposition of the Upper Cretaceous. Perm, Permian; Tr, Triassic; L, Lower Jurassic; BS+MJ+O, Middle Jurassic; Kim+P, Upper Jurassic; W, Lower Cretaceous; UK, Upper Cretaceous; T, Tertiary.

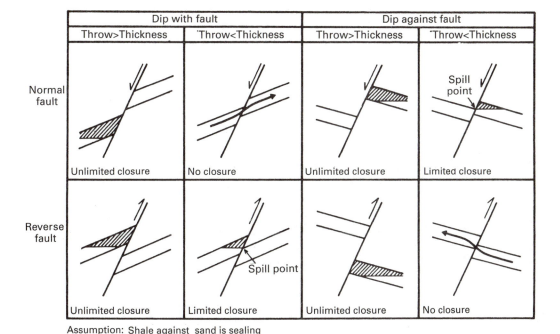

	Dip with fault		Dip against fault	
	Throw>Thickness	Throw<Thickness	Throw>Thickness	Throw<Thickness
Normal fault	Unlimited closure	No closure	Unlimited closure	Spill point Limited closure
Reverse fault	Unlimited closure	Spill point Limited closure	Unlimited closure	No closure

Assumption: Shale against sand is sealing
Sand against sand is non-sealing

Fig. 2.29 Six trapping and two non-trapping configurations against a fault, depending on whether the fault is normal or reverse, on the direction of dip of the beds relative to the fault plane, and on the amount of displacement of the reservoir. It is presumed that petroleum cannot escape up the fault plane.

or have acted in the past, in both ways. All very puzzling! Although attempts have been made to investigate the problem in Nigeria (Fig. 2.30) and elsewhere, and naturally we have some ideas on the subject, we still do not fully understand what the difference is due to. It adds further uncertainties to our predictions of the subsurface occurrence of oil and gas.

2.3.3 Stratigraphic traps

Petroleum may be trapped where the reservoir itself is cut off up-dip, thus preventing further migration; no other structural control is needed. The variety in size and shape of such traps is enormous, to a large extent reflecting the restricted environments in which the reservoir sands or carbonates were deposited. It would be pointless and boring for us to try to list all of the possible types of stratigraphic trap that can exist, so we will mention a few to convey the general idea, and leave the reader to speculate on other possibilities.

First, however, let us note that a number of traps, some of them very important, are formed by *unconformities* (described in Section 1.2.3); they differ somewhat in principle from the others, but are generally classified as stratigraphic traps. A dipping reservoir, cut across by erosion and later covered above the unconformity by impermeable sediments, provides the classic case (Fig. 2.21): the East Texas field, for example, is the biggest in the USA outside Alaska (it would be in Texas, of course!).

Unconformity traps can also be found above the break. Consider the sea gradually encroaching over the land as sea-level rises; the beach sands will spread progressively over the land surface, becoming younger as time goes on, until perhaps the supply of sand runs out. We would be left with a sandstone reservoir dying out above the unconformity, to provide a trap when later covered with, say, mudstone. More esoterically, but nevertheless known, a hill on the old land surface may be formed of permeable rock; if drowned by

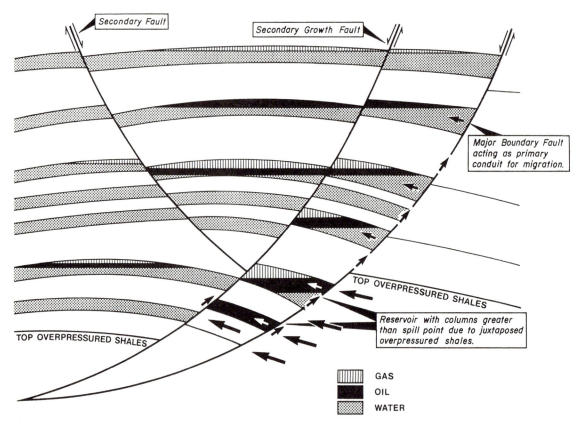

Secondary Fault

Secondary Growth Fault

Major Boundary Fault
acting as primary
conduit for migration.

TOP OVERPRESSURED SHALES

Reservoir with columns greater
than spill point due to juxtaposed
overpressured shales.

TOP OVERPRESSURED SHALES

GAS
OIL
WATER

Fig. 2.30 An investigation into the sealing qualities of faults affecting roll-over anticlines in the Niger Delta, where the reservoirs overlie overpressured shales. Where a reservoir is full to spill-point against a fault, and where an oil–water contact is continuous across a fault, it is presumed that the fault is non-sealing; elsewhere it appears to form a trap. The difference is believed to be due to clay being smeared into the fault plane, where there is enough of it in the section, as the fault moved. After K.J. Weber.

shales, the porosity could be preserved beneath the unconformity. Other situations can be imagined.

Non-unconformity traps are even more diverse. We mention just three examples. A coral reef may have grown as a mound on the sea-floor; if later overwhelmed by muds, we have an isolated stratigraphic trap (Fig. 2.21). A sand deposited in a river channel will be confined by the banks and, if terminated up-dip as not infrequently happens, again we have an isolated trapping situation (Fig. 2.21). A flood of sand washed off the shallow continental shelf into the deeper ocean, possibly through a submarine canyon, will spread out as a fan over the ocean floor (as illustrated in Fig. 1.10); its edges will provide an example of a reservoir dying out laterally. A lot of oil has been found in recent years in this sort of trap in the North Sea and elsewhere. In fact, fan sands provide one of the prime present-day exploration targets, although such prospects are not easy to locate and may require a lot of very sophisticated seismic. As the more easily found structural traps are running out in much of the world, there always seems to be something new to challenge our geologists and geophysicists.

2.3.4 Combination traps

A number of fields, some of them large, occur in traps formed by a combination of structural and stratigraphic circumstances; neither completely controls the trap. Again the range of possibilities is almost infinite. A couple of examples may give the idea.

The Prudhoe Bay field in northern Alaska (this *is* the biggest field in the USA!) has most of its oil and gas trapped in a Carboniferous to Jurassic sequence which includes more than one reservoir; these beds were folded into a faulted east–west anticline, tilted westwards, and truncated by erosion. The oil is held in the reservoirs by younger shales overlying the erosion surface (Fig. 2.31).

The oil in the Argyll and many other fields in the North Sea (Fig. 5.2) is trapped in tilted and faulted Permian to Jurassic reservoirs, which were eroded and unconformably overlain by Cretaceous shales. Both the faulting and the unconformity control the traps.

We may note here one most important consideration. The oil in these fields can only have migrated there *after* the traps were sealed by the higher sequences, or the oil would have been lost. This vital factor, that the trap must be shown to have been there before the oil migrated, possibly even before it was generated, is yet another aspect of the petroleum geology that we have to assess in proposing exploration drilling. The timing of trap formation versus oil migration has not always worked out favourably—to our chagrin.

2.3.5 Hydrodynamic traps

Take a look at Fig. 2.32. It shows water, perhaps from rain, entering a reservoir formation, or aquifer, up in the hills and percolating through it towards a spring off to the right. Oil has found its way into the reservoir and is battling to migrate up towards the surface against the flow of water. Depending on the balance of forces acting on the oil, it may find itself held against an unevenness of the reservoir surface where there is no conventional trap at all. This is what has been described as a *hydrodynamic trap*. It is totally dependent on the flow of water and is effective, of course, only for as long as the water keeps coming: dry up the supply of water, and the oil will be free to move again. This may be one of the reasons why oil accumulations trapped hydrodynamically are rare; a regime of water flow cannot normally be expected to remain constant for long, geologically speaking.

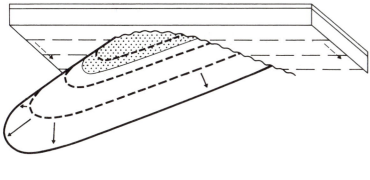

Fig. 2.31 A block representation of the trap at the Prudhoe Bay field in northern Alaska. The reservoir beds were folded into an anticline, which was tilted west and eroded before deposition of the overlying beds now dipping east. This *combination trap* is partly structural (the anticline) and partly stratigraphic (beneath the unconformity).

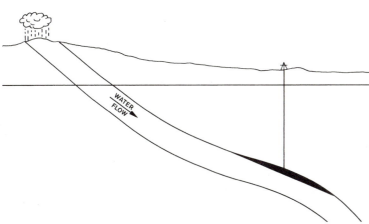

Fig. 2.32 A hydrodynamic trap. Oil, attempting to escape to surface up a reservoir, is held against an unevenness of its upper surface by water flowing in the opposite direction. There is no structural or stratigraphic closure. Note that the oil–water contact is tilted down in the direction of water flow.

Take another look at Fig. 2.32. You will notice that the oil–water contact is tilted down in the direction of the water flow. Such tilted contacts, in say ordinary anticlinal traps, are not all that rare; they are known in a number of parts of the world. In this sort of situation, we would have to be careful where we locate and drill our oil production wells in a field, as we do not want to waste the money drilling wells that would miss the oil altogether. Furthermore, cases are known where flowing water has apparently been able totally to flush oil out of an anticlinal trap. We would recognize this from residual traces of oil in a water-bearing reservoir, indicating the former presence of an oil accumulation now lost. It is therefore always important for us to try to get a handle on the hydrodynamic regime in a reservoir for both exploration and oilfield development purposes.

2.3.6   The relative importance of traps

A review by J.D. Moody in *Petroleum and global tectonics* (ed. A.D. Fisher and S. Judson. Princeton University Press, 1975) of 200 giant oilfields (those containing 500 million barrels or more) suggested the figures shown in Table 2.2.

These figures, although old, are revealing; they probably have not changed significantly since 1975. They emphasize the importance of structural, essentially anticlinal, traps in both number and size. The numbers of fields may partly reflect the fact that structural traps are easier to find than the others, but the reserves figures show clearly that generally they are also bigger. The trouble, from our present-day point of view, is that in most parts of the world the larger anticlines have now been drilled. What our efforts are increasingly directed towards, therefore, are the more obscure and generally smaller prospects. This is the challenge that demands our best science, our best technology, and our best interpretations.

Table 2.2

Type of trap	Number of fields (per cent)	Proportion of reserves (per cent)
Structural	66	78
Stratigraphic	22	13
Combination	12	9
Hydrodynamic	—	—

2.4   THE HABITAT OF PETROLEUM

We have now considered the geological factors that lead to accumulations of oil and gas in the subsurface. We have followed the story from the original organic remains, from which petroleum is formed, through to the places where we would seek to look for it. The essential key to successful exploration is that all parts of this story must come together, and in the correct sequence: if any one part has failed, then we cannot expect there to be any oil or gas there at all.

We can break the story down into five essential requirements, which we sometimes refer to informally as the 'magic five'. There must be:

1. A *source rock*. Generally a shale, or very fine-grained limestone, with a minimum of 0.5 per cent of the type of organic matter that will give rise to petroleum.

2. *Heat*. Obtained from the Earth by burial of the source rock, and required in order to generate petroleum from the organic matter. A temperature of approximately 150 °F is needed for oil to be generated; above about 350 °F only gas is produced; beyond some 450 °F even that is destroyed.

3. *Reservoir*. A layer or formation of rock that is both porous and permeable; usually sandstone or a carbonate.

4. *Cap rock* or *seal*. An impervious layer above the reservoir to retain the petroleum within it; usually a shale or evaporite. Sometimes the source rock itself may act as the seal if it directly overlies the reservoir.

5. *Trap*. A subsurface environment, formed by structural or stratigraphic control, where the petroleum in the reservoir is barred from further migration, and therefore accumulates. Note that the trap must have been there *before* the oil migrated. Note also that, once in the trap, the petroleum must be preserved there; later tilting or faulting could allow it to escape, or further deep burial could expose it to temperatures that could lead to its destruction.

To see these requirements fulfilled, we first have to home in on a sedimentary basin. A thickness of at least 7000–8000 feet of sediment will be needed to ensure that source rocks, if only at the base of the sequence, are mature to the oil generation threshold. The general positions of such basins around the world are shown on Fig. 2.33. It is no good our searching elsewhere,

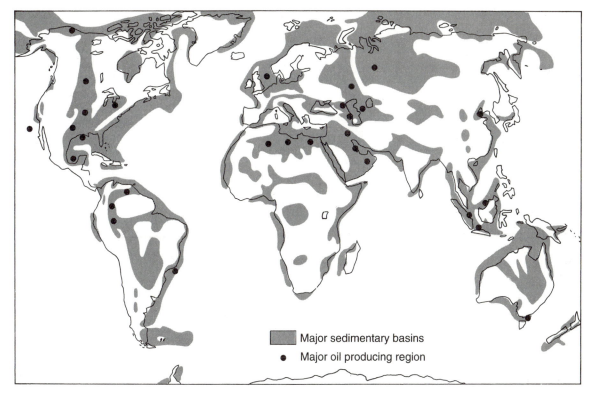

Fig. 2.33 The major sedimentary basins of the world. Those that are significant oil-producing regions are also indicated.

unless conceivably oil from a sedimentary basin could have migrated up to shallower depths.

The basic approach to exploration then is to consider separately and together each and every one of the 'magic five', to try to assure ourselves that each has been satisfied in our area, and then to see whether they come together in a satisfactory pattern. If we think they do, then we may be able to justify the cost of an exploration well; it not, then we had better go and try our luck elsewhere.

Of course it is not always as easy as this. There is likely to be a lot that we do not know about the geology of the subsurface, and this is where we have to start making our predictions (as explained in Section 1.1). We may be able to get considerable help by making comparisons and drawing analogies between our exploration area and other parts of the world that we know more about. We may be able to apply what we have seen and learnt elsewhere to a new and apparently similar geological setting. Consciously or subconsciously we use our knowledge and experience to help us in exploration judgements. It has been said that

the best geologist is the one who has seen most rocks, and that the best petroleum geologist is he/she who has seen most oilfields—please take both statements with a pinch of salt, although there is more than an element of truth in them.

This really is the purpose of this chapter: to see how the 'magic five' come together in different parts of the world to provide our petroleum resource. The ways that they do so are many and varied. In order to study them systematically, and to have a basis for making comparisons between different regions, we need to have a frame of reference for classifying sedimentary basins. This is conveniently provided by *plate tectonics*.

2.4.1 Plate tectonics

The principles of plate tectonics were formulated in the late 1960s after two decades of oceanographical work. In essence they are elegantly simple, and they provide us with a picture of how the Earth ticks.

The continuous ridges beneath the oceans, which more or less encircle the globe and of which the

Mid-Atlantic Ridge is a part, are formed by molten material (*magma*) welling up to surface from deep in the Earth. There it solidifies and, as more of it wells up, the solidified rock is gradually displaced away from the ridge crest. In this way new Earth's crust is continually being created.

The formation of new crust, if uncompensated, would result in a progressively expanding and less dense Earth; but we know that this is not happening, at least at anything like the rates demanded. To counterbalance the creation of new crust at the *spreading ridges*, older crust elsewhere disappears back down into the bowels of the Earth along so-called *subduction zones*. These are marked by deep ocean trenches bounded either by island arcs or by mountainous continental margins, in both of which volcanoes are characteristic (Fig. 2.34).

We thus have a picture of the Earth's crust being renewed along the spreading ridges, moving laterally away from them, and ultimately disappearing back into the interior along the subduction zones. Following this picture, the Earth's surface can be divided into a number of coherent segments or *plates*, each in motion relative to the others (Fig. 2.35). A plate is thus bounded by a *constructive margin* at a spreading ridge, where the horizontal stresses are extensional; a *destructive margin* at a subduction zone, where the crust is in a state of horizontal compression; and, at the sides, by *transform margins* where one plate slips sideways past another.

This is the essence of plate tectonics. It now provides us with an explanation for the forces that cause extension, as expressed perhaps by normal faults (where there is a tendency towards spreading); compression, evidenced by folding and reverse faulting (where two plates meet head on); and the subsidence of basins as warping due to crustal stresses. Actually there are other causes for basin subsidence; we shall meet some of them later on in this chapter. Side effects from these prime causes can become widespread and complex, but we need not now be surprised at the scale of the vertical and horizontal movements that have affected the Earth's surface.

However, we must take the story a stage further, and differentiate between the oceans and the continents. The crust in oceanic regions is relatively heavy, being formed directly from the material of the dense interior. The continents, on the other hand, are composed of generally lighter rocks derived ultimately from reconstituted sedimentary and segregated igneous rocks. The continents, therefore, are relatively buoyant, topographically higher, rigid blocks virtually 'floating' on the denser oceanic material, much as an iceberg floats in water.

During the movements of the Earth's crustal plates, the continents sit there as passive passengers carried along by the moving oceanic crust. This will continue until a continent eventually reaches a subduction zone where, if that subduction is located at an opposing continental margin, the two will meet head on. An almighty collision will take place, and the resulting compression will form a mountain belt. The ranges extending from the Alps, eastwards through the Balkans, Turkey, and Iran to the Himalayas resulted from such collisions during the Cainozoic era (the last 65 million years).

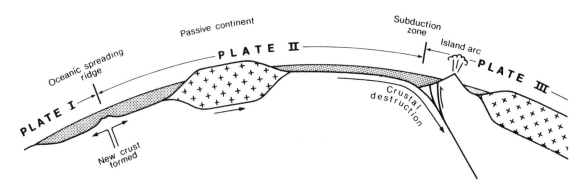

Fig. 2.34 The essence of plate tectonics. A plate is shown with a constructive margin at an oceanic spreading ridge, and a destructive margin at a subduction zone. Note that the continent is a passenger during the movement of the otherwise oceanic plate, and that it will eventually collide with another one across the subduction zone.

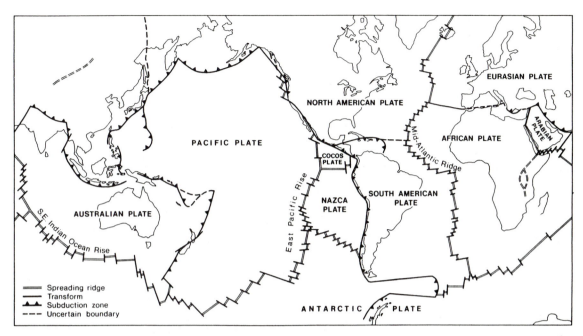

Fig. 2.35 The principal plates forming the Earth's crust at the present time; some minor ones are not shown. Note that most plates have a constructive and a destructive margin, some also have a transform margin. The African plate is the exception being surrounded on three sides by an oceanic spreading ridge: it is getting progressively larger.

Continents, being buoyant, are reluctant to go down into the dense interior. Consequently, following collision across a subduction zone, the whole system may become locked and the location of the subduction may shift. In western Canada, for instance, relatively small continental blocks are believed to have collided in the past with the margin of North America and to have been welded onto it; the subduction jumped further to the west. In the case of a major continental collision, the global pattern of plate margins may be altered. New spreading zones may be initiated, possibly through the middle of continents, thus splitting them in two; subduction zones may die and new ones be born.

The history of the Earth's surface is thus one of slow but continual evolution and change. We should not be daunted or surprised by the knowledge, say, that before about 120 million years ago the southern Atlantic Ocean did not exist and South America was joined to Africa (Fig. 2.36), or that India has moved from a position adjacent to Antarctica to its collision with central Asia. The rates of such *continental drift* are of the order of 2–10 cm per year, now becoming measurable over a number of years by precise satellite

position fixing, and having an enormous cumulative effect over geological time.

Continental drift also provides an explanation for the fact that most of the continents have passed through different climatic zones during their history. It is why, for example, we can find evidence that southeastern Arabia and Australia were glaciated during the Permian (some 250 million years ago), whilst northern Europe was in the desert latitudes that the Sahara is in today. It is the continents that have moved, not that the world was topsy-turvy.

2.4.2 The global distribution of petroleum

Plate tectonics gives us a useful framework within which to study and compare the sedimentary basins of the world, together with their contained oil and gas. However, as always, we have to be careful (the reader will be getting used to this by now!), because plate tectonics does not provide all of the controls on the types of sediment being deposited, and hence on the occurrence of petroleum. The climate, for example, will affect the nature of the sediment reaching the sea; and, as we have seen, oceanic conditions may govern the

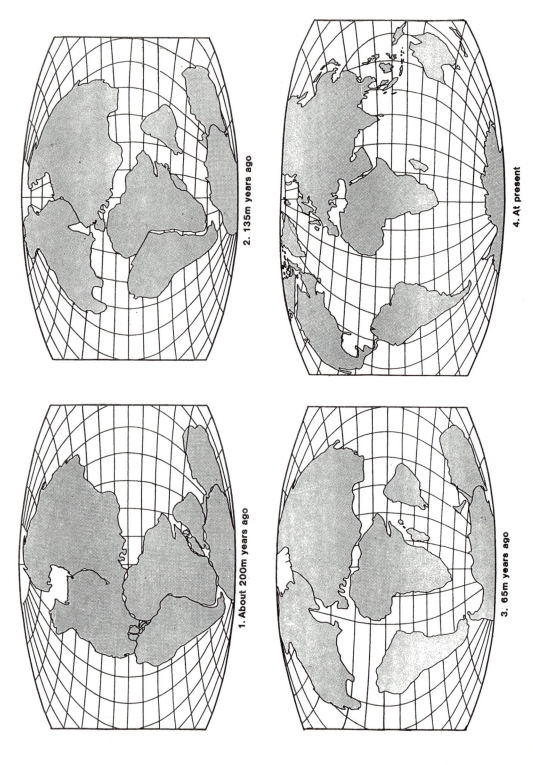

1. About 200m years ago

2. 135m years ago

3. 65m years ago

4. At present

Fig. 2.36 Continental drift: the changing pattern of relative positions and latitudes of the continents during the past 200 million years. Note that, in the Triassic, by coincidence they all came together to form one super-continent, known as Pangea.

abundance of marine life, and hence the richness of our source rocks. Nevertheless, we have got to have something to base a classification on and plate tectonics does more to help us than anything else.

Various attempts have been made to classify basins, with varying degrees of success. Nature is so variable that, if we try to be too sophisticated about the business, then each basin could fall into a category of its own; this clearly would not be helpful. So we prefer to keep things simple, and to consider basins associated with (Fig. 2.37):

1. crustal spreading;
2. crustal destruction;
3. transform plate margins;
4. basins inside continents.

Basins associated with constructive plate margins
By far the greater part of the world's petroleum occurs in basins which we can assign to a sequence of development, from an initial split through a continental mass, through a narrow seaway as the two sides start to drift apart, to a full-scale ocean, with a spreading ridge in the middle and stable continental margins on either side. We refer to this as the 'rift–drift sequence': the various stages of it are shown diagrammatically in Fig. 2.38.

We are not concerned with the spreading ridge itself, out in the middle of an ocean, as there is little or no supply of sediment here to provide us with either source rocks or reservoirs.

Imagine, however, a new spreading centre starting to develop through a major continent. As molten magma pushes up from deep in the Earth, the crust on top may be uplifted and tension will soon begin to stretch it. It will then fracture on normal faults, dropping down towards the axis of the stretching. This will lead to a trough along that axis, a *graben* or *rift valley*, which will start to be filled with sediment eroded off the uplifted shoulders of the rift.

This is the first category of sedimentary basin to be formed by the rift–drift sequence, but it is not going to be of great interest from the point of view of petroleum. The basins are small, there is not likely to be much in the way of source rock other, perhaps, than muds deposited in lakes like those of the East African rift, and there is not much opportunity for traps to be created. There is a further snag. Rift valleys are liable to be associated with volcanoes as the molten rock gets up near to the surface, and volcanoes can be bad news: the fine volcanic ash is liable to be mixed in with the sediments derived by erosion of the uplifted rift margins, and to clog up the pore spaces of potential reservoir sands. Although traces of oil have been found and there are a few small fields in the Rhine Graben of Germany, very little has been discovered worldwide and we cannot be really excited by the possibilities.

As the incipient spreading process continues, however, the rift is going to widen, with more faults developing on the flanks , and it will further subside. Eventually the sea will break in to fill the trough; this is the stage that the Gulf of Suez has reached. Now things start to look much more promising for us. The volcanoes have disappeared, but the sea may still receive some emanations from below through fractures, providing mineral nutrients for the marine life-chain, and hence possibly rich source rocks which are likely to be matured at relatively shallow depths. Waves and tides will wash and sort the sediment, still derived from the rift shoulders, leading to clean sands

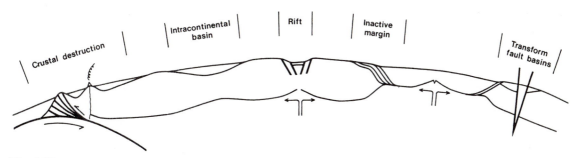

Fig. 2.37 The plate tectonic classification of sedimentary basins. Shown are the end members of the sequence of development from rift valley to open ocean, basins associated with crustal destruction and with transform faults, and those in the middle of continents.

Stage 1: Uplift and fracturing

Stage 2: Rifting

Stage 3: Further rifting; marine invasion

Stage 4: Initiation of new ocean

Stage 5: Ocean formation; development of inactive margins

Fig. 2.38 The 'rift–drift' sequence of extensional basin development. Please see text for full explanation.

to act as reservoirs when buried; and, in low latitudes, we may have a favourable environment for the growth of coral reefs, also providing future reservoirs. The sea-bottom sediments may blanket the tops of collapsed and tilted blocks between earlier but now dead faults, possibly creating combination traps of the sort that we illustrated in Section 2.3.4 by reference to the Argyll field in the North Sea. And lastly, if this sea has a restricted opening to the open ocean, we may find evaporites being formed; these may provide not only good cap rocks to underlying sediments but also, later after burial, additional traps (Fig. 2.26). So the geology is now much more

encouraging, and the Gulf of Suez itself contains several sizeable oilfields.

With continued stretching, the magma eventually makes it to the sea-floor and starts to form new oceanic crust in the centre of the rift: the process of genuine ocean spreading gets under way. The Red Sea illustrates the early stage, and the Atlantic may serve as our model for a full-blown ocean. The continental margins on both sides gradually stabilize, although they continue to cool and to subside gently for some time, creating possibly wide continental shelves and considerable thicknesses of sediment may accumulate on them (Fig. 2.39). Petroleum environments now will depend

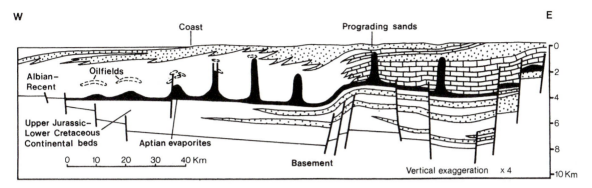

Fig. 2.39 Sketch representation of the African margin in Gabon, from the continent on the right to the Atlantic Ocean to the left. A layer of salt (black), which forms plugs, marks the initial invasion of the rift by the sea: it separates the underlying non-marine sediments of the rift valley, from the overlying marine deposits of the continental shelf at the edge of the widening ocean.

very much on local factors, such as the amount and nature of sediment reaching the sea; the climate and oceanography which may determine whether we have enough shelly animals to give limestones; and whether or not the processes are active to create any of the types of trap that we discussed in Section 2.3 (non-compressive anticlines, stratigraphic traps, etc). This is where it becomes dangerous to generalize too far, although we can still get guidance as to the types of sediment and traps that we could expect.

A special situation may be created off the mouths of major mud and sand carrying rivers, where a vast amount of sediment builds out into the edge of the ocean as a huge delta. The various environments of deltaic sand deposition (Fig. 1.10) provide the reservoirs, while the organic rich muds accumulating, say, between the river channels or offshore, may contribute to the source rocks. Common traps in this setting are roll-over anticlines (see Section 2.3.2). The Niger (Fig. 2.40) and Mississippi deltas are the prime examples, with large numbers of albeit rather small fields.

The North Sea is a basin belonging to the rift–drift category, but it is a rather peculiar one. It sometimes happens that the rift process gets under way but fails to continue to create a full ocean, becoming aborted usually at the Gulf of Suez stage, i.e. lots of faulting down towards a central rift or graben infilled with marine sediment. After all the activity has ceased, the basin may still subside gently as it cools, being topped up with an almost unfaulted sequence of strata. This is known as a 'failed rift' basin. In the North Sea basin, a succession of highly faulted Permian, Triassic, Jurassic, and Lower Cretaceous beds at depth, is over-

lain by a broad saucer of very little disturbed Upper Cretaceous and Tertiary sediments (Fig. 2.41). Oil and gas occur in both sequences (see Chapter 5). We may make comparisons between the sediments and structures of the lower sequence and, say, the Gulf of Suez, which were of considerable assistance in the early stages of North Sea exploration.

Finally, let us see what happens when an oil-bearing continental shelf eventually collides with another continent. This happened in the Middle East some 20 million years ago. A wide continental shelf on the edge of Arabia collided with central Iran, an ocean formerly between them having disappeared in the process. The outer part of that shelf was compressed and folded by the collision to form the huge anticlines in Iran which contain prolific oil derived from, and in part redistributed from, older shelf sediments. The inner parts of the shelf, in contrast, were preserved undeformed beneath the present Persian (or Arabian, depending on where you live) Gulf and eastern Arabia. The extremely abundant oil here is still in the original continental shelf structures: drape–compaction anticlines, salt domes, etc. (Fig. 2.42 and 2.43).

We have spread ourselves somewhat in our descriptions of the rift–drift sequence, because more than 80 per cent of the world's petroleum is found in basins representing one stage of it or another. They may also give some idea of the range of problems that the exploration geologist has to consider, and of the ways in which his/her thought processes function. We will now be briefer.

Basins at destructive margins At the opposite edge of a plate, where it encounters an opposing one, one

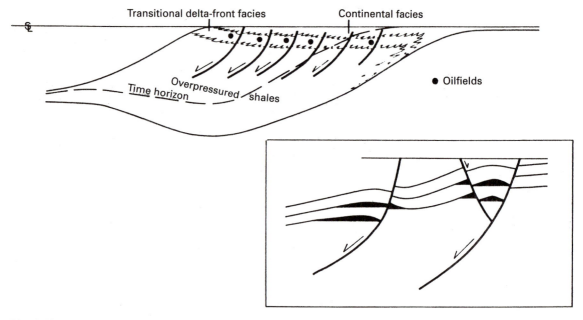

Fig. 2.40 Sketch cross-section of the Niger Delta, which has built out across the margin of Africa into the deep waters at the edge of the Atlantic Ocean. A series of faults flattening downwards into overpressured clays represents a sort of landslip down towards the ocean. The faults are responsible for roll-over anticlines which contain the oilfields (inset).

of the two plates is being subducted beneath the other. If one of the two plate margins is formed by a continent, it is the oceanic one that goes down. Conditions now are very different: overall the crust is in a state of compression, even though crustal warping may produce restricted tensional zones (Fig. 2.44).

Subduction is associated with steep and narrow continental margins, usually leading down to an oceanic trench; volcanism is characteristic. The sediments are rapidly deposited, variable laterally, and often mixed with volcanic ash which adversely affects their porosity and permeability. Structures are commonly very complex.

This does not amount to a very attractive picture for petroleum and, indeed, few commercial fields have been found in these environments. It is only in the less strongly compressed zones of destructive margins, for example behind the volcanic arc in Sumatra or in the Cook Inlet of southern Alaska, that significant oil and gas have been encountered.

There is an additional drawback. Such continental margins are commonly mountainous (e.g. the Andes or the Cordilleran ranges of North America), and operating conditions can be extremely rough. It could well become very expensive indeed to carry out exploration and to engineer production facilities. With a prospect of finding mostly small and complex oilfields, the costs may be unjustifiable. Offshore, the water depths may anyway be prohibitive.

Basins at transform margins   Small and complicated basins are associated, in land areas and beneath the continental shelf, with transform faults, at the sides of plates where one is slipping sideways past another. We would not normally think of giving them too much attention were it not for the fact that some of those in California are amongst the richest in the world, in terms of barrels of oil per cubic mile of sediment.

Some 35 separated small basins lie on- and offshore on both sides of the renowned and dreaded San Andreas Fault, which slices through California from north of San Francisco to the Mexican border (Fig. 2.45). They have evolved very rapidly, mostly during the last 20 million years, but they contain a considerable number of oilfields in often very complex structures. The reason they are there is nothing to do with plate tectonics; it is the richness of some of the source rocks. A region of deep oceanic upwelling lies

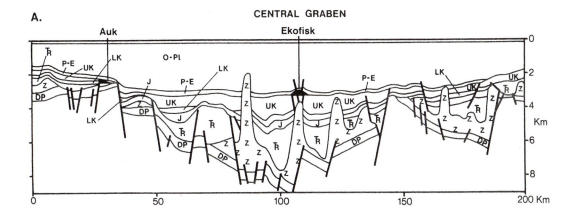

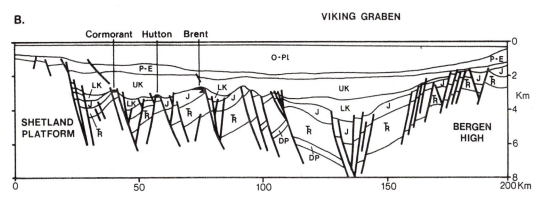

Fig. 2.41 A failed rift basin: the North Sea. Profiles across the central and northern North Sea, showing the lower rift-faulted sequence, with the faults mostly inclined towards the centre of the rift (graben), and the younger post-rifting infill of the basin. DP, Devonian to Permian; Z, Permian salt (central North Sea); Tr, Triassic; J, Jurassic; LK, Lower Cretaceous (during which faulting mostly ceased); UK, Upper Cretaceous; P-E, Paleocene-Eocene; O-Pl, Oligocene to Pliocene. Some oilfields are named. After P.A. Ziegler.

off the Californian coast and this nutrient-rich water supports the whole marine life-chain in great abundance—hence the prolific source rocks, and some fine seafood restaurants.

There is always liable to be something to upset the generalizations of the petroleum geologist. By this time, you might be beginning to think that he/she could be feeling somewhat beaten around the head—but perhaps the Californian seafood provides some compensation!

Basins within continents A number of sedimentary basins lie entirely within the major continents. In some cases, they appear to owe their origin to incipient rifting and to be failed rifts, but others evidently do not; we just do not know what has caused them to subside. They are commonly very long lived, maybe for up to 500 million years, and are little affected by folding or faulting.

Some of these, the Williston Basin in north central North America (Fig. 2.46) for example, are oil-bearing; others are less so or not at all. The difference seems to depend, at least in part, on whether or not the sea was ever able to penetrate the basin... but let us refrain from further wild and unjustified generalizations!

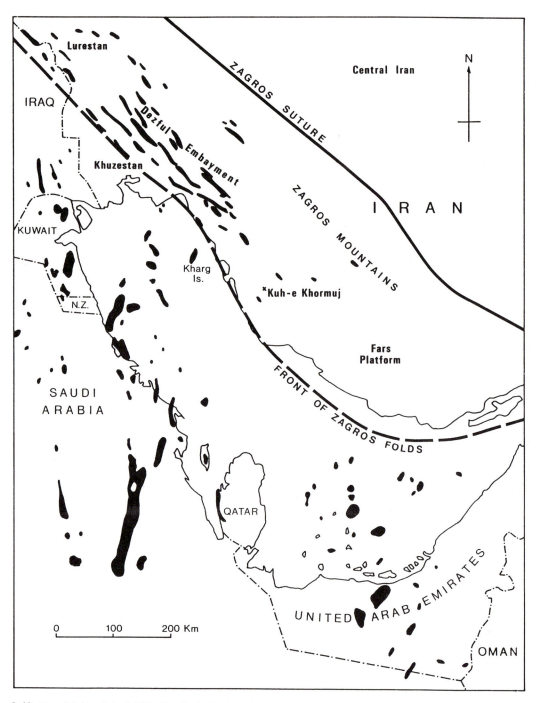

Fig. 2.42 The oilfields of the Middle East basin. In the Miocene the Afro-Arabian continent collided with central Iran along the Zagros Suture and back-pressure folded the Zagros Mountains but this has not affected the area of the Gulf and Arabia. The north-west–south-east orientated fields in Iran are in Zagros anticlines; in Arabia those elongated more or less north–south are in drape–compaction anticlines, whilst the more circular ones are in salt domes.

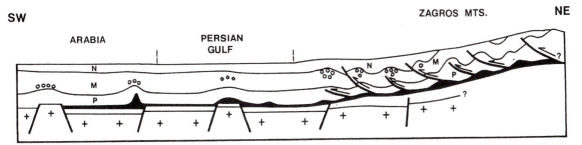

Fig. 2.43 Profile across the Middle East basin from the Zagros Suture (right-hand side) to Arabia (left). The positions of the oilfields are shown, as explained in Fig. 2.42. N, Neogene molassic sequence including evaporites; M, Mesozoic/Palaeogene shelf sequence; P, Palaeozoic; Black shading, Eo–Cambrian salt; +, continental basement; circles, petroleum accumulations.

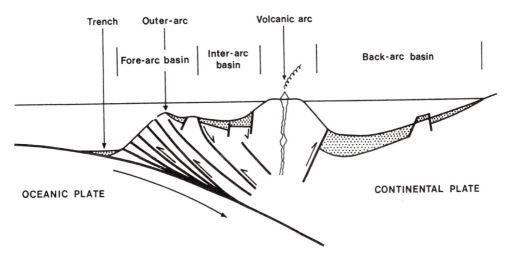

Fig. 2.44 The general positions of small, complex basins associated with subduction zones. Little petroleum has been found in these environments, except locally in the back-arc basin.

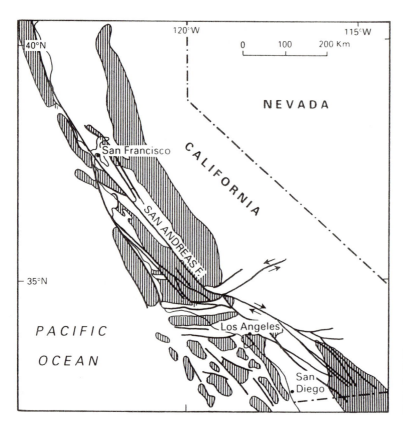

Fig. 2.45 The many small, complex basins (vertical hatching) associated with the transform San Andreas Fault, and its earlier branches, in southern California.

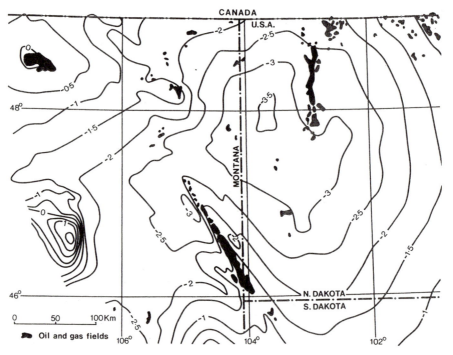

Fig. 2.46 The shape of the Williston Basin in North America. The contours are on a Cretaceous horizon in thousands of feet below sea-level and major oil and gas fields, located on very gentle anticlines, are shown. After T.P. Harding and J.D. Lowell.

3

Exploration methods

3.1 SURFACE TECHNIQUES

We saw in Section 2.4 that exploration in principle means studying the 'magic five', and trying to convince ourselves that they are all satisfied in a correct association in our area. If so, then there should be petroleum down there; if not, then there won't be. It sounds easy, but we can never be one hundred per cent certain of these factors down under the ground where we cannot see the rocks. However much information we have, there will always be some uncertainty in our understanding; all we can do is to arrive at the best evaluations possible from the information available to us. It is the means of acquiring such information, as accurately as possible, that are outlined in this chapter.

3.1.1 Preliminary study

Before we undertake commitments or start to spend money in an area that we may be considering for exploration, we would naturally first find out all that we can about it. Only then would we decide whether or not to apply for an exploration licence and whether to spend money to obtain more data.

At this early stage, there are essentially two sources of information: the *published literature* (and any unpublished material that we might be able to buy or otherwise honestly lay our hands on!), and *remote sensing* which may be available for purchase.

Geological investigations, already carried out for scientific purposes as well as in the search for oil, have over the years uncovered a vast amount of information. Of course more is known about some regions than others. At the least we could expect to find out whether our area contains a sedimentary basin at all; at the best, there may be enough known already to enable us to make a fair initial evaluation of the undiscovered petroleum possibilities, and hence to decide whether to go ahead with an exploration programme. Information is available primarily through the geological literature, but a lot of it regrettably has never been published and

may be tucked away in obscure corners. Our first job then must be to lay hands on whatever we can find: our library staff can be worth their weight in gold in this!

Remote sensing takes two forms: satellite imagery and photography from aircraft. *Satellite images*, which are not photographs strictly speaking but rather scanned views like television pictures, can be recorded, processed and displayed in a variety of ways designed to bring out features of interest, in our case the rocks at the Earth's surface (Figs 3.1 and 3.2). They can be printed out using a range of colour schemes and at any desired scale. Thus, we may see a whole country displayed on a single picture, or we may look at detail as small as some 10 m: different rocks and their structures may show up quite clearly. We may thus be able to learn quite a lot about the geology of a little known region. Perhaps the greatest value of such images, however, is that they can provide a grand overview of much, or even the whole, of a sedimentary basin, enabling us to spot and follow large-scale features that might otherwise have escaped our notice. And, of course, they comprise excellent maps for a wide variety of uses. All of the world has been covered many times, and appropriate images are readily available.

Aerial photographs are normally taken systematically looking vertically downwards, but with sufficient overlap between successive pictures so that they can be looked at stereoscopically to make hills and other features stand out in 3D. They have been taken on a range of scales, and can often reveal considerably more detailed information on the geology than do the satellite images. However, being taken from aircraft, rather than orbiting satellites, their availability is controlled by national governments and, for some countries, they may be impossible to obtain. They have considerable use for mapping, and hence for strategic purposes.

Thus, from the results of previous work and from remotely sensed pictures, we may be able to obtain a lot of knowledge of a new area without ever stirring

Fig. 3.1 Mosaic of Landsat images of southern Iran. The geology of the upper right two-thirds of the picture area is clearly different from that to the south towards the Gulf, where several elongated anticlines are visible. Data available from the US Department of the Interior, US Geological Survey EROS Data Center.

from our offices—and it is cheap! However, sooner or later, either before or after we have obtained a concession or licence to explore there, we shall need to go to our area of interest to collect more detailed information by one or more of the techniques described below.

3.1.2 Geological survey

The traditional method of exploring for oil onshore was to map the surface geology and then to extrapolate downwards to the subsurface. It often took the petroleum geologist to remote parts of the world, and could be both adventurous and arduous physically (Figs 3.3 and 3.4). Perhaps the most exciting and personally rewarding period of petroleum history was the first half of the twentieth century, when companies were extending their activities to remote and often unknown parts of the world. Geologists were in the vanguard and the tales of the old-timers can be both entertaining and hair-raising!

Times, alas, have changed. Much exploration has moved offshore where no rocks are visible and, on land, enough is now known about the surface of most sedimentary basins to provide the answers that we

would seek, at least initially. There is still, however, a place for surface geological study, to assist us in extrapolating into the subsurface, and to study in detail aspects of the geology that enable us better to interpret the patchy information that we can obtain from the subsurface. Regrettably, few young petroleum geologists these days can expect to be able to spend much of their time in the field; but dare one older geologist suggest that both they and their companies might sometimes be more effective if they could?

What we can learn from careful mapping, measuring, and studying the surface outcrops will help us in understanding the geology of other, and maybe unexposed, parts of the basin and at deeper levels. Such things include:

1. The thickness of the succession in at least one part of our basin. Measuring the thicknesses of successive beds will help us not only in such geological exercises as predicting the organic maturity of various beds, but also in knowing what orders of depth we may have to drill to.

2. The ages (from fossils) and nature of the sediments in the basin. Do we have potential source and reser-

Fig. 3.2 Part of a satellite image of south-west Iran. The grid crosses are 10 km apart. A syncline is clearly seen in the upper centre of the picture with, to the south, the beds curving round the plunging end of an anticline. Different rock types are responsible for the different topographic textures visible. Photograph from the Department of Geology, Imperial College.

voir beds? How thick are they? How rich are the source rocks, and how porous and permeable (at surface) are the reservoirs? Whereabouts in the succession do they occur?

3. The environments of deposition of the various sedimentary layers. This knowledge will help us in reconstructing past palaeogeographies and hence in predicting the distribution of source and reservoir in the subsurface.

4. The structure of the beds. Are there anticlines that might offer traps at deeper levels? Is the basin affected by faults and, if so, are they normal or reverse? Are unconformities developed and, if so, at what levels? When were these various structures formed?

5. Are there any natural seepages of oil or gas in the area? Petroleum migrating underground will escape to surface, if it is not trapped *en route*. There it is lost by evaporation or by degradation (Section 2.1.2). Seepages tell us a lot: that there must be a mature source rock down below, that the oil has been expelled from the source rock to accumulate somewhere, and that the accumulation is leaking. What seepages do not tell us is precisely where that accumulation is, or anything about its size; it could well be too small to be worth exploring for. As an aside, we may note here that techniques have been developed, and are used by some companies, to detect microseeps both onshore at the ground surface and offshore from the air; they show that there is at least some petroleum in the area, and this

Fig. 3.3 A geological field survey camp in the foothills of the Brooks Range, northern Alaska. The Carboniferous rocks forming the mountain have been thrust up and over considerably younger beds surrounding the lake.

Fig. 3.4 A geologist sampling an outcrop in northern Alaska. The gun is not normal equipment: it was carried here, but not used, as a last-ditch protection against bears.

may provide the encouragement needed to spend more money on detailed exploration.

It may be a fairly straightforward matter to get all this information if the rocks are well exposed at surface, for example along the coast or in the mountains of arid regions. If, however, the land is covered by soil or by thick vegetation, we are likely to see very little, unless we can make use of artificial exposures such as quarries or roadside cuttings: it is in such circumstances that clues obtained through remote sensing can be of considerable help. In the last resort, we may have to dig small holes or trenches, or to drill shallow wells either by hand or by a small vehicle-mounted rig, to collect samples of the solid rock.

Samples that are collected and brought back from the field, together with those collected from wells (see Section 3.3.1), may be subjected to a whole range of analyses, carried out by specialists or specialist groups and all designed to tell us something more about the rocks, their history, and their relationships to petroleum. To mention but a few, palaeontologists and palynologists will tell us the ages of the rocks and something about their environments of deposition; petrographers will report on the composition of the rocks and the changes that they have undergone since deposition (diagenesis); geochemists will study the nature and maturity both of the source rocks and of the oils themselves; petrophysicists will be concerned with porosity and permeability. Making and interpreting these analyses may become businesses in themselves; and their results then have to be incorporated into our assessments of the petroleum geology. They all add to the vast spectrum of information that affects the outcome.

However we get it, every little piece of information is going to help us to build up the three-dimensional picture of the geology of our area, and of the possible distribution of hydrocarbons. We need all the help that we can get!

3.2 GEOPHYSICAL TECHNIQUES

Geophysical techniques enable us indirectly to sense aspects of the geology of the subsurface. They involve the measurement of some physical quantity related to the nature of the rocks and their distribution; interpretation of the results then gives us information on the geology. The most important of them to petroleum exploration are the 'potential field' (gravity and magnetic) and seismic methods. Others, such as electrical resistivity surveys, are less useful to us and will not be considered here.

3.2.1 Gravity surveys

Very small changes in '*g*', the acceleration due to gravity, are caused by variations in density of the rock types under the ground. Heavy rocks at depth will cause '*g*' to be higher than it will be over less dense ones.

A gravity survey involves the precise measurement of '*g*' at close stations along a track (a road perhaps), using a very sensitive instrument known as a *gravimeter* (Fig. 3.5). Positive deviations (*anomalies*) from the regional average suggest heavy rocks close beneath the surface, as we might find in the core of an anticline, whereas negative anomalies imply a relatively light rock below, such as salt. A grid of such survey lines will allow the anomalies to be mapped out across the area, and for calculations to be made on them to give us clues to the underlying rock types and structures.

The gravity technique is usually regarded as a reconnaissance tool, which is useful particularly in the preliminary stages of an exploration programme. It was, for example, spectacularly successful in locating salt domes and plugs during the early exploration of the US Gulf Coast region. It can be used both on land and at sea, and the measurements can be made from the air, even from satellites.

3.2.2 Magnetic surveys

Rocks also have different magnetic properties. Those that have a relatively high natural magnetism cause

Fig. 3.5 A gravity survey in western England. The squatting geophysicist is reading the gravimeter and the theodolite is for precise position fixing. Photograph by Royal Dutch Shell.

minute local effects on the intensity and direction of the Earth's magnetic field. Somewhat similarly to gravity variations, these effects can be measured and mapped to show the distribution of magnetic anomalies over the area of interest. Surveys are most commonly conducted from the air, by means of a magnetometer carried by an aircraft flying along set lines of survey.

The relative strength and the sharpness of the magnetic anomalies is controlled by the depth below surface of magnetic rocks, particularly of the igneous and metamorphic rocks forming crystalline basement. The prime use, therefore, of a magnetic survey is to indicate the depth to basement, i.e. the total thickness of sediments in the sedimentary basin, and to locate the major structural features within it.

3.2.3   Seismic reflection surveys

Far and away the most important geophysical method is the seismic reflection technique: some would say that it is the most important contributor to the whole of exploration. Indeed, seismic is used at various stages of an exploration programme, and especially before drilling to ensure that a well is located in the optimum place.

In principle, the method is simple: we are merely echo-sounding on the various sedimentary layers in our basin. However, so valuable is the information obtained that extreme efforts are made to ensure that it is as clear and as accurate as we can possibly get. These efforts are such that the method now commands an extensive hi-tech industry in itself, involving a wide range of specialist expertise and very expensive equipment. Most of this section must therefore be devoted to it.

Background: the general idea   If we make a sharp noise of any sort, the sound energy travels out in all directions as a series of moving vibrations. Most of it gets lost but, if there is a wall say in the way, a small part of the total energy is reflected back to us as an echo; and another small part of the energy passes on through the wall. The further away the wall is from us, the longer it takes for us to hear the echo. We can use this idea to calculate the speed at which the sound travels; or, conversely, if we know the speed of sound, we can calculate how far away the wall is. This is how reflection seismics work. We make a noise at the surface of the ground, or of the sea, and its energy spreads away in all directions although we try as much as possible to direct it downwards. When the fraction of it that does go straight down encounters a particu-

larly hard layer underneath a soft one, or vice versa, some of it is reflected back to surface; most of it, however, carries on downwards and, again, some will be reflected back from the next hard layer. Up at the surface we shall hear, or detect, a series of echos representing the reflections from the successive bed boundaries. The times, after the initial shock, of arrival of successive reflections, and their strengths, can be tape recorded, and this record can be played out visually so that the reflections stand out as a series of kicks or wiggles. Then, if we know the speed (velocity) of travel through the overlying rocks, we can calculate the depths to the reflecting horizons.

If, now, we repeat the process at a series of stations along a survey line, we can arrange the successive records beside each other so that the kicks line up to give us a two-dimensional picture of the rock layers, looking for all the world like a geological cross-section (Fig. 3.6). Actually it is not a cross-section because the 'depth' is usually plotted in time in seconds: the two-way travel time (TWT) is the time it takes the energy to go down from the source to the reflecting layer and back up again. This is what is measured and it is fact; to convert it to actual depth in feet or metres involves knowing the velocity through the overlying beds, which is likely to be very variable, and means that 'depth conversion' may be highly interpretative. If next we map out the travel times to a reflector at many points on a whole grid of survey lines, we end up with a map analogous to a structure contour map but contoured in seconds of TWT.

To try to improve the efficiency of the operation, more than one receiver (known on land as a *geophone* and offshore as a *hydrophone*) is used. Several are planted along the survey line, in both directions from the source onshore and in the direction behind the boat offshore, and the recordings are made simultaneously; these days the recording is usually digital. This means that we obtain reflections from several points on our reflecting horizons at the same time. Such points are spaced at half the distance between the source of energy and the corresponding geophones at surface (Fig. 3.7).

Noticing this enables geophysicists to be even more cunning. It will be appreciated that the actual percentage of the energy of the source that comes back in a reflection is really very small, and it is often difficult to separate out the genuine signal from a mass of background noise. We can try to improve on this, by repeating the process with a geophone now placed where the source was the first time. This will give us further reflections from the same spots on the subsur-

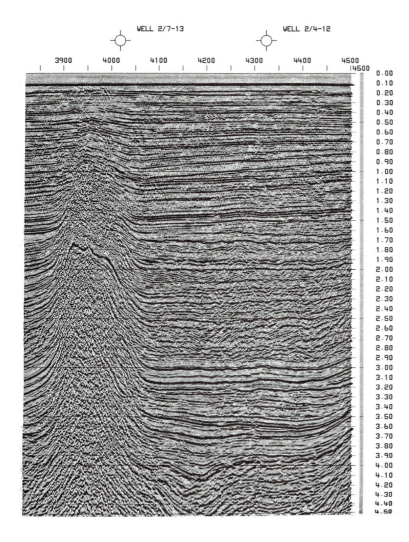

WELL 2/7-13   WELL 2/4-12

3900 4000 4100 4200 4300 4400 4500 |4500

0.00
0.10
0.20
0.30
0.40
0.50
0.60
0.70
0.80
0.90
1.00
1.10
1.20
1.30
1.40
1.50
1.60
1.70
1.80
1.90
2.00
2.10
2.20
2.30
2.40
2.50
2.60
2.70
2.80
2.90
3.00
3.10
3.20
3.30
3.40
3.50
3.60
3.70
3.80
3.90
4.00
4.10
4.20
4.30
4.40
4.50

Fig. 3.6 Part of a seismic profile from the Norwegian North Sea. The length of the profile is approximately 25 km and the vertical scale at right is in seconds of two-way travel time. Note that the profile is compiled by setting beside each other adjacent records; the tops of these records can be seen at the top, above a strong reflection from the sea-floor. The near-horizontal reflections represent the boundaries between rock layers of differing composition, and the feature on the left is interpreted as a salt-plug. Profile from Nopec.

face reflector that we covered before (CMPs or common mid points) (Fig. 3.8). If then the two records are added together, we should get double the signal strength and hopefully cancel out some of the noise. The process can be extended to give over a hundred fold coverage of all CMPs.

This is the general principle of the seismic reflection technique. Let us now look at some aspects of it in a little more detail to appreciate what it is that our geophysical colleagues are doing.

Data acquisition The techniques of obtaining the basic seismic data differ onshore and offshore, although the general principles are the same: we need a source for the noise or energy that is put into the ground, and a means of detecting and recording the returning reflections.

Onshore. One of the most efficient sources of energy is a charge of dynamite, exploded in a shallow well drilled by a small hand or truck-mounted rig (a shot-hole) (Fig. 3.9). This is still used in remote areas but it is not very acceptable near civilization. One still refers, however, to the seismic shot and to the shot-point—the location of the shot which is surveyed very carefully. The geophones are laid out (planted) on either side of the shot-point along the line of survey: they consist of small microphones mounted on spikes that can be driven into the ground. The signals are collected and recorded, usually digitally, at a recording station (Figs 3.10 and 3.11) or in a movable cabin.

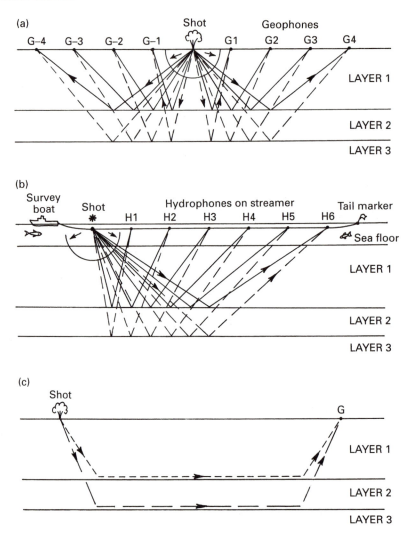

Fig. 3.7 The principles of seismic survey. (a) Seismic reflection survey on land: the geophones are spread out evenly in both directions from the shot along the survey line, to receive energy returned to surface from points on the reflecting horizons mid-way between shot and geophone. (b) Marine reflection survey: the hydrophones are evenly spaced along a cable towed behind the boat and behind the energy source. (c) Seismic refraction survey: the shot is distant from the geophones, so that some of the energy recorded will have travelled along the rock layer boundaries themselves.

Various other ways of creating seismic energy have been used, as for example dropping a heavy weight onto the ground. Most common, however, is the 'vibroseis' technique. A steel plate mounted under a heavily weighted truck is lowered to the ground and the back of the truck lifted so that most of its weight is carried by the plate. This is then vibrated, so that the ground is shaken. It is these vibrations that are reflected back from depth to the geophones. The vibrations have a particular pattern, or signature, that can be recognized by the recorder enabling it to measure the TWT. More shake may be obtained by using more than one vibroseis truck, and it is an impressive sight to see a family of them in line, all vibrating away (Fig. 3.12). The technique is effective, it creates little disturbance and it causes no damage to the environment or even to a road surface.

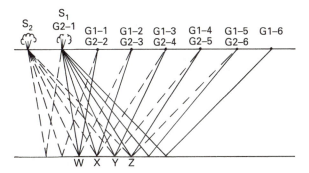

Fig. 3.8 The principle of multicover reflection survey in a marine configuration. A reflection from, say, common mid point X will be received in hydrophone G1-2 from shot-point S1, and in hydrophone G2-4 from S2. The two records may be added together to enhance the reflection signal.

Offshore Marine seismic tends to be quicker and cheaper since much of the equipment can be towed behind a boat, and recording can be virtually continuous. The delays of moving from station to station and of redeploying the geophones each time are avoided.

Again, dynamite has largely been superceded as a source—the fish did not like it very much! Nowadays, an air- or gas-gun is used, in which a chamber is either explosively voided of compressed air or else evacuated and implosively filled with gas. More than one such gun may be used simultaneously to enhance the signal (Fig. 3.13). The guns are towed behind the seismic boat and can be fired at intervals of a few seconds.

The hydrophones are arranged along a cable, or streamer, which may hold as many as 240 individual receivers along a length of several kilometres. It is towed behind the guns and is held a few feet below the surface of the sea, the tail being marked so that other shipping knows where it is. So that records can be obtained from common mid points, the timing of the shots is adjusted to the speed of the boat and to the spacing between the hydrophones on the cable.

Once all of the individual records have been printed out and arranged side by side, the result of such shooting and recording along a straight line will be a seismic section or profile (Fig. 3.6), and a grid of such lines will enable our contour map to be drawn; the accuracy required will determine the spacing of the lines, or the size of the grid squares.

It will be appreciated that the ability to conduct such a survey is dependent upon very precise position fixing. Positions at sea have to be, and can be, determined to a few metres using both radio beacons and satellites.

Even more dependent upon knowing exactly where we are is the so-called 3D survey, which in effect consists of a series of swathes of very closely spaced lines. One way this can be achieved is to tow a gun from either side of the boat (or from two boats) and behind them to arrange four parallel streamers (Fig. 3.14). The shots from each gun will be picked up in the

Fig. 3.9 A seismic shot-hole being drilled in Indonesia. Photograph by Royal Dutch Shell.

Fig. 3.10 Seismic recording equipment being assembled in the mangrove swamps of Nigeria.

Fig. 3.11 A seismic survey recording station in Nigeria. Photograph by British Petroleum.

hydrophones in all of the streamers. Thus some of the energy being received will have travelled oblique to the direction of the boat and, when compounded together, the records give a blanket coverage across the width of the swathe: we are surveying an elongated area, rather than a single line. When the boat turns around and steams back, the adjacent swathe will be blanketed, and so on until the entire survey area is covered. The number of records and the amount of information that has to be handled is mind-boggling, and it may take weeks of computer time to process the data from such a survey. Such, however, is the value of this really close-spaced information, that 3D surveys are becoming standard practice in regions like the North Sea, not only for oilfield development purposes but for exploration.

Data processing   The raw records, or traces, that might be played out from the geophone are scarcely usable as they are. They have to be combined and treated in various ways to bring out the information that they contain. Now that the data can be recorded digitally, these operations are managed by computer; and therefore a lot of information and complex manipulations can be handled. This is the processing that the data are subjected to. The precise form of the steps followed will vary depending on the nature and quality of the data, and perhaps on the location of the survey; they can be adjusted to bring out the best results. Interactively controlled processing has resulted in

Fig. 3.12 A crew of four vibroseis trucks in Oman, with the vibrating plates lowered onto the ground. Photograph by Royal Dutch Shell.

Fig. 3.13 An array of four airguns being lowered into the sea from a marine seismic survey boat. Photograph by Royal Dutch Shell.

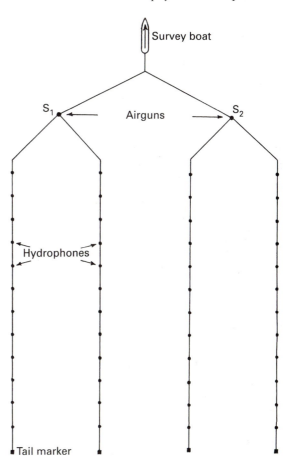

Fig. 3.14 One possible arrangement of airguns and streamers for a marine 3D seismic survey. Reflected energy from both sources will be received in all hydrophones. Other arrangements may be used, employing, for example, two boats.

enormous improvements in output quality over the past few years.

Some of the things that we are trying to achieve are as follows.

First of all we have to get the individual traces lined up properly. For example, a reflection will take longer to reach a distant geophone than a nearer one; if we just lined up the travel times, a really flat reflecting horizon would appear to get deeper and deeper towards the more distant geophones (Fig. 3.15). The effect is known as 'normal move-out' (NMO) and must be corrected for.

Onshore, we will have to allow for the fact that the ground surface is rarely completely level; differing geophone elevations have to be compensated for through the application of 'static corrections'.

Secondly, we have to add together the data from the different records of each common mid point, i.e. they have to be 'stacked'. The idea is that this addition will

enhance the genuine reflections but will cancel out much of the background noise. Other processing steps, involving a degree of digital juggling, are designed further to amplify and sharpen up the information that we want, and to eliminate or minimize what we do not.

Thirdly, there are unwanted effects which have to be got rid of. It may be that some of the reflected energy, on its way back up towards the surface, gets turned back down again from the bottom of a higher layer, to be reflected from the original horizon a second time (Fig. 3.16(a)). This will give us the appearance of a reflection where we do not really have one at all, a so-called 'multiple'. In an extreme case, the energy may bounce to and fro several times between two layers to

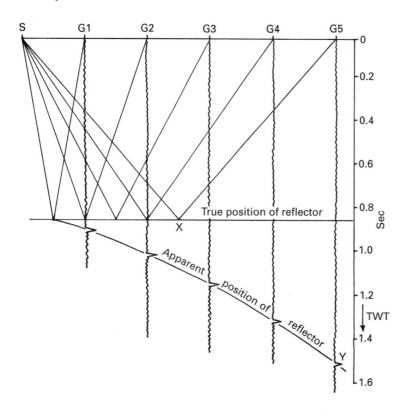

Fig. 3.15 The effect of normal move-out (NMO). Because of the greater horizontal distance travelled by energy received in the further geophones, a relection from X appears in G5 as if it were from Y with a spurious two-way travel time. This effect is corrected for during processing.

give a series of multiples; and similar bouncing around between the sea-floor and the surface can create strong sea-floor multiples (Fig. 3.16(b)). These can all be confusing to the real reflections, and we want to process them out.

Another unwanted effect is caused by the energy being scattered from sharp corners underground, such as might be provided where a hard bed is chopped off by a fault (Fig. 3.16(c)). This can give rise to curved patterns on the seismic profile, known as 'diffractions'. They can sometimes be useful in helping us to identify and locate small faults, but equally they can obscure the real reflections and it is generally desirable to try to get rid of them.

Fourthly, we may need to 'depth convert' the data. This involves knowing the velocities of all the beds being surveyed, i.e. the speed with which the sound travels through them. It will vary, even for a single formation, depending on lateral variations in the rocks and on the depth of burial and the degree of compaction of the rock. The velocities can be estimated from the NMO corrections (referred to above) or, better, measured directly in a well through the

beds—if there is one. This is done by lowering a string of geophones into the well and comparing the times that energy from a shot at surface takes to reach them.

Finally if the beds in the subsurface are dipping, the returning energy that is received by a geophone actually comes from a point that is not half way between shot-point and geophone (Fig. 3.16(d)). We build it into the seismic profile, however, as if it came from a common mid point, i.e. in the wrong place. This will give a false picture of the geology, and alter the apparent dip of the beds. The effect can be corrected for during the processing; this is known as 'migration'—but is nothing at all to do with the migration of petroleum! When we look at a seismic profile, it is important to check to see whether or not the data have been migrated.

All of these steps, and others, mean that the processing of seismic data can be long, complex, and expensive. It is big business.

Seismic interpretation   We have collected our data and processed it. Now we have to see what it is telling us.

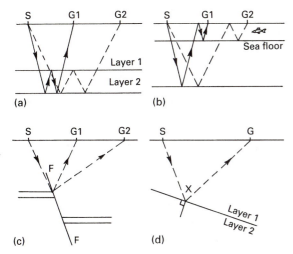

Fig. 3.16 Some problems encountered in seismic reflection survey. (a) Multiples: in addition to normal reflections received by the geophones from a point on the base of Layer 2 mid-way between S and G, we may have energy following the paths shown, being reflected twice from that horizon; this will produce the appearance of a second, deeper reflection. (b) Energy following these paths may give rise to a sea-floor multiple. (c) Energy scattered from a point such as the faulted termination of a reflecting horizon may be received by successive geophones to give an apparent reflection (a diffraction pattern) cross-cutting the true reflections. (d) The reflection from a dipping reflecting horizon will not be from a point mid-way between shot and geophone, although it may be plotted out as if it were: the effect must be corrected by 'migration' of the data. Attempts are made during processing to correct for or deal with all of these problems.

Traditionally, the geophysicist collected together all the profiles of the area, picked out particularly prominent reflections with coloured pencils (Figs 3.17 and 3.18), followed them through the grid of lines, plotted their depths (in seconds of TWT) on maps of the pattern of lines, and contoured around the depths to the reflectors at each shot-point. The resultant maps could be viewed as structure contour maps, once they had been depth converted and once the reflections were identified as coming from particular beds. It is convenient that, quite commonly, strong reflections are obtained from the top of reservoir formations that we are especially interested in. Thus we are able to identify possible structural traps and potential drilling locations.

Unfortunately, none of this interpretation is easy and much of it is subjective. However, help is at hand!

These days, the picking may be done on an interactive workstation, and differing interpretations of the data can be experimented with until the most satisfactory one (but not necessarily the correct one!!) is arrived at. The results can be displayed not only to show the depths of the reflectors but also the varying strengths, or other attributes, of the reflected signals; these also give information on various properties of the rocks. We may end up with a whole series of coloured images, each of which tells us something different about the geology.

Yet another way of looking at the results is on a 'time slice', a display of the reflections cutting any given depth level. Thus we may see a 'map' of the geology at a constant depth of, say, 1 s of TWT which gives us an alternative way of looking at what the rock layers are doing at depth.

Lastly, let us note that seismic can help us in different ways altogether. The strength and character of a reflection tells us something about the nature of the rocks themselves, and the patterns of the main reflections, as well as of minor ones between them, give clues as to the environments of deposition of the sediments. This study has become known as *seismic stratigraphy*. Furthermore, just occasionally the seismic gives us direct clues to the presence of petroleum, in particular if it is gas: the so-called '*direct hydrocarbon indicators*' (DHIs). For example, there may be sufficient contrast between the properties of a sandstone containing gas in the pore spaces and one containing water, to cause a seismic reflection from a gas–water contact: a 'flat spot'. These DHIs, however, are not unambiguous, they can be caused by other things, and there will still be some uncertainty as to whether there is actually petroleum down there or not.

With all its manifold benefits, it is not surprising that the reflection seismic method is such a powerful tool for exploration, and that so much time, effort, and money are devoted to it.

3.2.4 Seismic refraction surveys

A seismic technique which is still used for reconnaissance purposes, but which used to be more common, is the refraction method. Here, the geophones are much further away from the shot, which therefore needs to be a powerful dynamite blast, and the energy actually travels for some distance along the top of the hard bed (Fig. 3.7(c)). This gives information over a wide area, but it tends to iron out the unevennesses of the structure

Fig. 3.17 A geophysicist interpreting a seismic profile by hand. Photograph by Royal Dutch Shell.

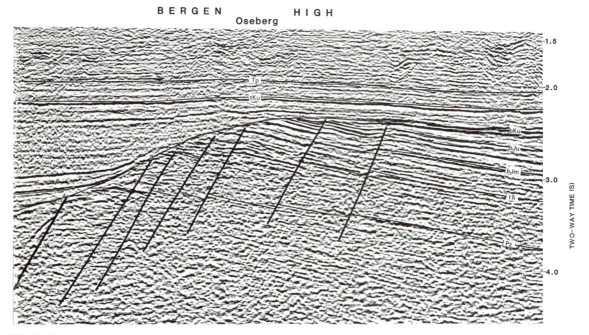

Fig. 3.18 Part of an interpreted seismic profile from the Norwegian North Sea. The length of the profile is about 15 km and the vertical scale is in seconds of two-way travel time. Note that the shallow part of the profile has been cut off. An unconformity is shown between east-dipping faulted beds below and nearly horizontal beds above. Pz, top Palaeozoic; tTr, top Triassic; bJm, base Middle Jurassic; bJu, base Upper Jurassic; bKu, base Upper Cretaceous; tKu, top Upper Cretaceous; Tp, a Paleocene horizon. Profile from Nopec.

rather than defining them; it is more difficult, therefore, to interpret in detail. We need not concern ourselves further with it here.

3.3 WELL DRILLING AND LOGGING

3.3.1 Well drilling

Most pictures of oilfield activity show a drilling derrick. This is what stands over the well being drilled and its function is to lower the string of steel drill-pipe carrying the bit into the hole, and to draw it out again. It is just the most obvious part of a complex set of machinery designed to drill a well to depths up to 5 or 6 miles as efficiently and as safely as possible (Figs 3.19 and 3.20).

Fig. 3.20 A land rig drilling at Wytch Farm, southern England. The travelling block carrying the drill-string is visible in the derrick, at the foot of which the derrick floor is shielded. The rig motors and mud tanks are on the right, and lengths of casing are stacked in the foreground ready for use in the well. Photograph by British Petroleum.

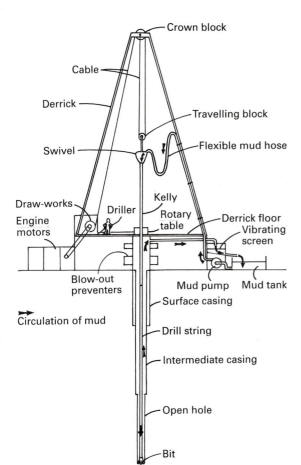

Fig. 3.19 The essential parts of the drilling rig. For explanation, please see text.

The 'cutting edge' of this system is the drill *bit*, the form of which depends on the nature and hardness of the rocks to be drilled. For most ordinary sedimentary rocks, the bit will consist of sharp teeth set into three somewhat angle mounted cones (Fig. 3.21). As the bit rotates on the bottom of the hole, so these cones go round and the rows of teeth cut into the rock. For harder formations, the bit may instead of teeth carry studs to crush the rock. If extremely hard rocks are to be drilled, the bit may consist of diamonds set, not on cones but into the head of the bit itself, and which scratch away at the rock as the bit rotates.

The bit is screwed into the bottom of the *drill string*, consisting of 30-ft lengths of steel pipe, the bottom few of which are particularly heavy and known as *drill collars*. The weight of most of the drill string is held in the derrick above ground by a cable passing over pulleys and through the *travelling block* to a massive horizontal winch called the *draw-works*; by means of this the pipe can be raised and lowered in the well and the amount of its weight resting on the bit itself can be

Fig. 3.21 A new bit being lowered into a well. Note the cutting teeth on the three rotatable cones on the bit. Photograph by British Petroleum.

Fig. 3.22 Driller controlling the drilling operation. The four-sided kelly passes through and is rotated by the kelly bushing, which in turn sits in the underlying rotary table. Photograph by British Petroleum.

adjusted. The bit, and the whole of the drill string, are rotated in the well by virtue of the top joint, known as the *kelly*, being hexagonal or square in cross-section and passing through a hexagonal or square hole in a rotating table at surface: the *rotary table*, holding a *kelly bushing* from which depths are usually measured (Fig. 3.22).

As the well is drilled the rotary table is turned by the rig motor, the kelly slips down through it, being rotated all the time and itself turning the drill string and bit below. When the length of the kelly is drilled down, it is unscrewed and another 30-ft length of drill pipe is added below it. When the bit gets blunt or needs changing, all of the drill string has to be pulled out of the hole, usually in 90-ft stands of three 'singles', to get at the bit. Such a 'round trip' has to be made at intervals usually of one to five days.

From time to time, it is necessary to protect and line the well bore by steel *casing*. Thirty-foot lengths of strengthened steel pipe are screwed together, lowered into the hole and cemented in position. Of course, once this has been done, the well has to be drilled further with a smaller bit that will pass through the casing, and the next string of casing will be smaller still. As there is a limited range of bit and casing sizes, the well has to be engineered carefully beforehand to make sure that we can drill down to the depths required.

A most important part of the equipment is the mud system. Carefully controlled liquid mud is pumped into the top of the kelly, down the whole of the drill string, out through nozzles in the bit, back up to surface via the space between the pipe and the wall of the well, through a series of sieves and into tanks from which it is recycled (Fig. 3.19). This mud fulfils several functions. It lubricates and cools the bit; the weight of the column of mud in the well holds back in the formations any pressured gas, oil, or water that they might contain; and it cleans out the hole, carrying the chips or *cuttings* up from the bottom. The cuttings are sieved out of the mud before it is recycled and, when they are cleaned, provide us with our first small samples of the rocks that are being drilled: they are routinely collected, examined (Fig. 3.23), and logged, together with other factors such as the rate at which the rocks are being drilled. The mud is also monitored for any traces of petroleum that it might have picked up at depth—perhaps the first indication that we may have made a discovery. The result is the invaluable 'mudlog' for which the data may be collected and compiled by specialist service companies.

If more information is required than can be obtained from the small cuttings, a core may be cut. For this

Fig. 3.23 A well-site geologist examining cuttings at a well in Holland. Photograph by British Petroleum.

purpose, the 'bit' has a hole in the middle so that, as drilling proceeds, a cylinder of rock moves up into the 30- or 60-ft 'core barrel' at the bottom of the drill string. This provides a much larger sample of the rock, from which perhaps sedimentary structures can be recorded, fossils collected, and measurements of porosity and permeability made. Despite the value of the information gained, coring has to be restricted as much as possible, since it is slow, entails round trips before and after cutting, and is therefore expensive.

Not all wells go vertically downwards. For a variety of reasons, it may be desirable or necessary to drill at an angle. Both the direction and angle of a deviated or directional well can be controlled very precisely, and the position of the bit monitored. In some cases it may be helpful even to drill virtually horizontally, to follow along a particular reservoir formation rather than to cut across it. The technique of horizontal drilling has been developed particularly in the past few years and is now fairly commonplace. One such well may take the place of several vertical ones, and may make it possible to develop and produce a field that otherwise would be uneconomic.

Lastly, let us mention an important safety aspect. Normally, the weight of the drilling mud is adjusted to hold the natural fluids back in the rocks and to prevent them escaping up the well. If, however, we unexpect-

edly encounter very high pressures, the gas, oil, or water might be able to flow into the well and to push out some or all of the mud at surface. In this situation the well might 'blow wild' with mud, gas, oil, and water all over the place—obviously undesirable. A series of clamping devices (*blow-out preventers* or BOPs) is therefore bolted onto the top string of casing in the hole, beneath the rotary table, as a second safe-guard against any such dangerous, wasteful, and possibly environmentally damaging eventuality (Fig. 3.24).

The equipment to do all of these things is referred to collectively as the rig. For land drilling it is broken down into units that can be transported by road or, in rough terrain, by helicopter. Everything is moved to the selected site and assembled, the well is drilled and, a month or two later when drilling is complete, it is moved off again and the land rehabilitated so that one would not be able to see that the rig had been there.

Offshore, everything has to be packaged onto as small a platform as possible, together with living accommodation, fuel tanks, etc. The unit has to be robust enough to withstand the severest sea and weather conditions. Modern offshore drilling rigs are masterpieces of design and engineering. There are essentially three types of rig: a jack-up in which the legs of the unit stand on the sea-floor and which is

Fig. 3.24 Blow-out preventers bolted onto the top of the casing, and sited beneath the derrick floor, on an offshore well, Gulf of Mexico. Photograph by British Petroleum.

usable in water depths up to some 300 feet; the semi-submersible, with the platform mounted on pontoons, which float deeply in the water to provide stability, and which needs anchoring in position (Fig. 3.25); and the drill-ship, which really is a ship with a central opening (moon pool) through which the hole is drilled, and which is often held in place by dynamic positioning (Fig. 3.26); it can drill in really deep waters. Finally, let us note here that, if the exploration well discovers a field that will be developed for production, then a much larger platform will be required with provision for several wells to be deviated from it and carrying a mass of production equipment as well; the size and complexity of such platforms are quite staggering.

3.3.2   Well logging

A well, that was expensive to drill, is largely wasted unless we extract from it all of the geological information that we possibly can; it would otherwise be little more than a mere hole in the ground. On the other hand, a well that is comprehensively logged provides the prime source of information from which to evaluate the subsurface petroleum potential; it also provides precise control for the interpretation of our seismic data.

Fig. 3.26 A drill-ship anchored off Ireland. The well is drilled through a hole in the centre of the ship. Photograph by British Petroleum.

There are essentially two methods of extracting information from a well. A preliminary log of the rocks penetrated may be built up from the study of the cuttings whilst the well is being drilled, and this is usually augmented by the drilling performance, for example the rate at which the bit can penetrate the rocks. Such a *mudlog* will be supplemented by detailed descriptions of cores, if we are fortunate enough to have any. The second technique is by the use of *wireline logs*, in which physical properties of the rocks are recorded by means of devices lowered into the well on an electric cable.

The mudlog   As described above, the cuttings from the bottom of the well, carried to the surface by the circulating mud, are sieved out and cleaned so that the well-site geologist has small samples of the rocks being drilled. These are collected, examined, and described at regular intervals. Their depths, however, have to be adjusted because the bit will have drilled deeper during the time that it takes them to travel to surface. They also get spread out and mixed during this time. The first log thus tends to be rather imprecise, but it can be corrected to true depths by incorporating the depths to rock boundaries from the drilling performance which is also recorded on the log.

The mud coming back to surface additionally is monitored continuously for traces of hydrocarbons. This may give us a first indication that we have made a discovery.

Fig. 3.25 A semi-submersible drilling rig off Ireland. Some lengths of drill-pipe are stacked in the derrick. Photograph by British Petroleum.

All this work is normally carried out by specialized contractors, often referred to unglamourously as mud-loggers. In recent years, they have extended their techniques to use a form of down-hole recording, akin to wireline logging (see below), to enable a log to be compiled as the bit is actually drilling the rocks. This *measurement while drilling (MWD)* today provides an important guide to the drilling process.

Cores, if we have them, after they have been examined and sampled for indications of oil or gas, are cleaned up and described to a degree of detail that is not possible from cuttings. They are particularly valuable in obtaining precise information to build up our understanding of reservoir formations, as described in Section 4.2. Direct laboratory measurements are also made, from cylindrical plugs cut from the cores, of the all-important porosity and permeability.

Wireline logging These geophysical techniques are also carried out by specialized contractors, the pioneer of which was the French–American firm Schlumberger; indeed the records are commonly referred to improperly as Schlumberger logs.

A device known as a *sonde*, containing measuring instruments, is lowered into the well on a coaxial cable. As it is slowly withdrawn up the hole, continuous measurements of certain physical properties of the rocks are transmitted electrically to surface, where they are recorded (Fig. 3.27 and 3.28). The records may be plotted on film (Fig. 3.29), or else recorded digitally for computer based interpretation.

Various combinations of physical properties may be measured by the same sonde, but it is commonly necessary to make more than one run to obtain all of the different logs needed. Most of the tools will not work through casing, and so separate logging surveys have to be made before each of the successive strings of casing is cemented into the well.

How these surveys work may perhaps best be illustrated by describing some of them briefly.

The *gamma ray log* plots the natural radioactivity of the rocks. A form of geiger counter in the sonde measures the trace quantities of radioactivity given off by most sediments; these quantities are minute, generally less than the radioactivity that surrounds us every day, and are certainly not at all dangerous. Some rocks, especially shales, contain more radioactive minerals than others, such as sandstones or carbonates, and the log recorded as the sonde is withdrawn from the well thus gives us a plot which reflects the rock type. The gamma ray log is therefore basic to the full suite of

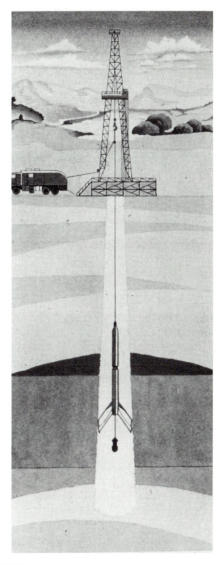

Fig. 3.27 The principal elements of wireline logging. The sonde, held by a cable from the logging truck, is centralized in the borehole. Photograph by Schlumberger.

logs normally run. A more recent development is to separate out the wavelengths of the emanations from uranium, thorium, and potassium, which all contribute to the total radioactivity of the rock; this can give us more detailed information about the composition of the rocks being surveyed.

A series of devices measure the electrical *resistivity* of the rocks. An electric current is sent out from the

Fig. 3.28 Cut-away diagram of a wireline logging truck. The sonde, being assembled in the foreground, is connected by a cable passing over a winch near the back of the truck, whilst the recording cabin is towards the front. Photograph by Schlumberger.

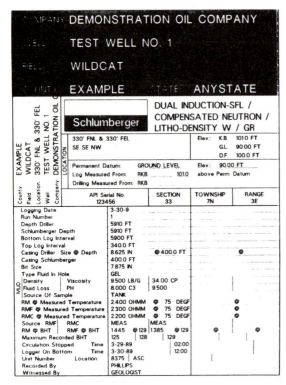

Fig. 3.29 The form of the header to a wireline log, recording the relevant well data at the time of logging. Photograph by Schlumberger.

sonde into the rocks surrounding the well and is picked up through another electrode; the voltage drop is measured and hence the resistivity is calculated. Some rocks are more resistive than others, so again we have an indication of rock type. But the resistivity logs do more for us than this. An electric current passes more easily through a porous rock filled with salty water than through one containing fresh water, and more readily through one filled with oil than one with gas. Thus the resistivity logs (Fig. 3.30) enable us also to estimate porosity and, further, to know whether it is filled with water, oil, or gas.

Another one, the *sonic* or *acoustic log*, records the speed of sound through the rock layers. Sound from an emitter at one end of the sonde travels through the borehole wall and is picked up by a receiver at the other end; the time lag is recorded. This not only provides a valuable control on seismic velocities, but also information on the nature of the rocks, because sound travels faster through some than through others. It also provides a measure of porosity since the sound waves take longer to pass through, say, a porous sandstone than a non-porous one: this provides an independent check on the porosity determined from the resistivity logs.

Yet another device is the *neutron log*. A stream of neutrons is sent out into the well bore from the sonde; they react with the hydrogen in the rocks generating gamma rays, which are then measured in the sonde. Since most of the hydrogen is contained in water or

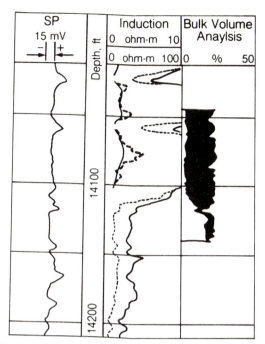

Fig. 3.30 A section of a basic electric log through a reservoir sand. The plot on the left is the spontaneous potential (relative voltage), and the electrical resistivity, measured by two devices, is in the centre. An interpretation of the oil saturation is on the right. Note that the well cuts through the oil–water contact at a depth of about 14 120 feet. Photograph by Schlumberger.

hydrocarbons, the strength of the signal again gives an indication of porosity and helps us to recognize gas as opposed to oil or water. This tool proves to be particularly useful in carbonates.

There are several other logs as well, and new ones are continually being developed. Combinations of many or all of them give us a detailed picture of the petroleum geology of the strata drilled through. What we aim primarily to determine are:

1. the rock types, or lithologies;
2. the porosity of potential reservoir formations;
3. the nature of the fluids contained in the porosity.

Note that permeability cannot be measured directly (yet!), but it can be estimated approximately from certain combinations of logs, especially when evaluated by computer programs.

It is through these wireline logs that much of the information is obtained on all wells that are drilled in the search for oil and gas. They form the basis of the standard well completion log. Where further and greater detail is needed then we have to take cores, but these of course cannot be obtained after we have drilled and logged the well. The geologist either has to anticipate the requirement in advance of drilling, or to cut the cores in a subsequent well. A partial solution may be provided by *side-wall samples*. These are small cylinders of rock cut out of the borehole wall, by a very hard steel, hollow 'bullet' literally fired sideways out of a special sonde known as a gun. They do provide us with little samples of the rock in place, but are generally too shattered for meaningful measurements of porosity and permeability to be made.

3.4   PROSPECT IDENTIFICATION

We have considered the sort of information that we are able to get from the subsurface, and the ways in which it is obtained. Now we must look at the use that is made of these data in the search for further accumulations of petroleum. How are the various pieces of information combined to make decisions about whether to explore or not?

In detail, the ways in which this work is organized may vary from company to company, although in principle they are all undertaking the same sort of evaluation. The process may be regarded as twofold: first, to build all of the information we have into as comprehensive as possible a picture, or model, of the subsurface geology, layer by layer, bed by bed; and then, secondly, to use this model to predict where we might expect (hope) to encounter oil or gas by drilling.

3.4.1   Geological modelling

The first step is to identify and recognize the successive layers that comprise the sequence of sediments in our area, be it a small licence block or an entire sedimentary basin. If any part of the succession of strata comes up to surface outcrop on land, then we can collect the fossils to date the beds, log the nature of the rocks, and measure their thicknesses. We are concerned to identify in particular the potential source and reservoir formations. If, as is commonly the case offshore, there are no outcrops, then we shall be entirely dependent on any wells that have already been drilled, taking our information from the logs, cores, and cuttings.

To make sure that we are following and mapping out the individual beds correctly, we shall need to

resource and that we are very concerned with the quantities involved. Now we must see how we can apply our knowledge of the geology to assessing the amounts of petroleum that we have found, or hope to find. Don't panic, we are not going to get ourselves involved with lots of numbers or mathematics! This section is included to give a further idea of the very difficult problems that we have to try to tackle. We will refer to oil, but the same considerations, methods, and terms can be used equally for gas.

First, let us again emphasize that we are dealing all the time with uncertainties. There is no way of knowing in advance of drilling whether or not there is going to be any oil or gas at all down there under the ground, let alone how much. And yet our companies need to know what we expect, or hope, to find. Similarly, once we have made a discovery, there is no way that we can know precisely how much we have found: the geology, which controls the amounts of oil in the reservoir, is liable to change between our information points, our wells. We have to try to understand, or predict, just what these changes amount to. So, until we have actually produced all of the oil that we ever shall, we are involved with a greater or less degree of uncertainty. How do we handle these problems?

Before we get into this, we have to be quite clear what it is we are talking about when we refer to quantities of petroleum. There is a good deal of misunderstanding and misuse, even within oil companies, of the following terms:

Oil in place: This is the total volume of oil, measured in barrels or other units, that is present in an accumulation under the ground. It usually refers to what was there originally, before we started to take any of it out. You may see the engineers using the term *STOOIP*: stock tank oil originally in place. The stock tank is, in the case of small fields, located at surface near the well-head, and oil may be produced directly into it; and hence the STOOIP refers to the oil in place in the reservoir but corrected to the volume it would occupy under surface pressure and temperature, and therefore without any dissolved gas of significance. We cannot regard these quantities as 'reserves', since we are never able to recover *all* of the oil that is down there in the reservoir.

Recoverable reserves: The volume of oil that can actually be produced to surface from an accumulation. We may distinguish between primary reserves that can

be produced without any artificial assistance other than pumping; secondary reserves which can be produced using assisted or enhanced recovery techniques (see Section 4.4); and tertiary reserves using more exotic techniques (also mentioned in Section 4.4). Note, however, that the proportion of the oil in place that we can recover, will depend on the economics—how much money we are prepared to spend on getting it out of the ground. A bald figure for 'recoverable reserves' is somewhat meaningless, unless we can be more specific about how we are going to produce them.

Because anyway there is uncertainty about the figure, it is desirable to be able to express our degree of confidence in it. This may be done via a standard deviation (the ± figure that we sometimes see after a number) or by a statistical probability (see below).

Proven reserves: Here we start to enter a minefield! Different companies have different definitions of what is proven. Some might use the term to refer to the amount of recoverable oil that is believed to lie within a given radius, half a mile or whatever, of a well; what they think is beyond that in the accumulation, they might designate 'probable'. Increasingly these days, companies tend to use 'proven' for those reserves that they believe to be present with an 85 or maybe 90 per cent degree of confidence or statistical probability. What this means and how we arrive at the figure, we shall see shortly.

Probable reserves: Equally dodgy! One definition was given above: the term may be used, like 'proven', to refer to a degree of confidence or probability, in this case 50 per cent. Sometimes 'possible' is also seen, to cover the reserves that have only a 15 or 10 per cent chance of being present. It may well be that it is best to avoid the terms 'proven', 'probable', and 'possible' altogether, and just to qualify our figures by statistical probabilities: at least then people would know what we mean! It is not difficult. Even the weather forecaster these days sometimes specifies, say, a 10 per cent chance of it raining this afternoon; we are talking about exactly the same sort of thing.

Original and *remaining reserves*: These are fairly obvious. They refer respectively to what was there and recoverable before we started producing, and what is still there for the taking at a given date. Usually, if we hear simply about 'reserves', it is the latter, the remaining reserves, that is meant.

Undiscovered reserves: A figure for what we *hope* to find in a prospect, area or sedimentary basin in the future. This figure is extremely imprecise and may be not much more than a guess (see below); we can still qualify it by a statistical probability, however. Adding this to the original reserves will give us what is sometimes called the '*ultimate reserves*'—a grand total for the basin.

Perhaps these explanations will give the patient reader some idea of what we are up against when we come to consider quantities of the resource on which a good deal of our civilization depends. It is far from easy, and in the end it all comes back to the geology.

3.5.1 Discovered reserves

Once a discovery of oil has been made, the normal way of estimating how much we have found is to start with the volume of the reservoir within the closure of the trap. We then eliminate progressively everything from this volume that is not oil. So we multiply the bulk volume of the reservoir in the trap by those factors that represent the non-oil:

Recoverable reserves =

$$\frac{BV \times Fill \times N/G \times \phi \times (1 - Sw)}{FVF} \times RF \times Constant$$

where:

BV is the volume of the reservoir formation within the closure of the trap above the spill-point. It is governed by the shape of the trap, faulting, and the thickness of the reservoir. BV will be determined from seismic and well data, and regional and local geological interpretation.

Fill is the 'fill factor', which is the percentage of the bulk volume that actually contains the oil, the volume of the gas cap and the water-bearing rock below the oil–water contact being discounted. It is affected by many factors, including the adequacy of the source rock to provide enough oil to the trap (this amount is sometimes called the *charge*), and the quality and strength of the cap rock. If we do not know where the gas–oil and oil–water contacts are, then this factor may be little more than a guess; if we do, then we can go straight to the bulk reservoir volume containing the oil.

N/G is the net to gross ratio. Not all of a reservoir formation is going to be sufficiently porous and permeable to contribute oil to production. We have to discount those parts of it that are useless and just consider the net reservoir thickness. This will be controlled by variations in the nature of the sediments that comprise the reservoir, meaning that we have to try to interpret in detail the environments that the sediments were deposited in. This can be pretty subjective, even when we have information from a lot of wells. What anyway should we regard as net reservoir? A rather arbitrary porosity cut-off value is often used.

ϕ is the porosity, or rather the average porosity of the net reservoir across the entire accumulation: a tall order! We do our best from measurements on core samples and from wireline log interpretation, but what happens between and beyond our well control?

Sw is the water saturation, the percentage of the porosity that is occupied by the immovable water. Again we need an average value for the field. We have not only all the problems of average porosity but remember that the size of the pores comes in here as well: the finer the sand, the higher will be the water saturation.

FVF is the so-called formation volume factor. This reflects the fact that oil under the ground in the reservoir occupies more space than it does when we get it up to the surface; it shrinks because gas bubbles out of it as its pressure is eased during production. We may actually be able to measure the FVF if we have a sample of oil collected under subsurface pressures from the bottom of our well.

RF is the recovery factor, the proportion of the oil in the reservoir that we can actually recover and produce. In a sandstone reservoir, this is commonly about 50–60 per cent, but it may be a good deal less from carbonates. It is a figure that we cannot know exactly until we have finished producing. So we usually have to base our estimate on prior experience elsewhere.

A constant is needed to adjust the units. The Americans measure reservoir volume in acre-feet: area in acres multiplied by reservoir thickness in feet. To get an answer to our sum in barrels of oil, we have to multiply the figure we calculate by 7758. If we are working entirely in the metric system, then we don't have to worry.

It will be clear to anyone who has got this far through the book that, in producing figures for all of these factors, there must be considerable uncertainty—to say the least. What we are doing, then, is to multiply uncertainties by uncertainties, doubtful estimates by doubtful estimates, until we begin to wonder whether our answer has any reality or meaning at all. Different geologists will certainly come up with different values for at least some of the input factors, and arrive at perhaps wildly different answers. Who is right? Whose answer should we use? Can we indeed believe any of them? Is the whole thing a load of codswallop? Unfortunately we cannot escape from the problem; our companies, consumers, and governments *must* have numbers that they can use for planning purposes, even though they may be well aware that any such figures will eventually turn out to be wrong.

We seem to have got ourselves into a right old mess. But we cannot just sit back and say 'Too bad!'; we have to do something. Most commonly these days, and to try to be as honest and objective as possible, the problem is tackled through a statistical technique (again, don't panic), known as a *Monte Carlo simulation*.

Instead of estimating single figures for the factors that go into the reserves formula (average porosity and the rest), for each of them we work out our best estimate, having regard to all of the geology; and we also specify the total range, from minimum possible to maximum possible, somewhere within which the 'true' figure must lie. Then we get a computer to pick a value for each factor at random from the range we have given, but biasing its pick towards our best estimate. The computer does the sum using these values. Then we ask it to do the same thing again, and again, and again... maybe 500 or 1000 times. So we have a whole list of answers, any one of which could be the real value. The list is put into order from the smallest to the largest, and then analysed statistically.

If we plot out the numbers, or percentages, of answers on our list falling within successive size ranges (in barrels of oil), we shall find that the bulk of them tend to cluster round the middle (Fig. 3.32). The one that has the most answers in we can regard as the most probable value (for the statistically inclined, this is the modal class of the distribution)—in other words, our best estimate, i.e. what we are trying to express. More commonly, however, we give as our preferred figure the average of all the answers (the mean): add them all together and divide by the number of them—the computer does this very easily. This is because, for

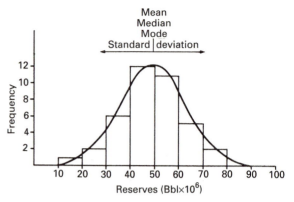

1. Normal distribution

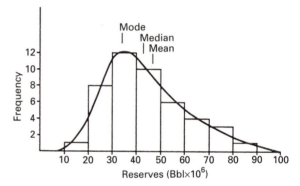

2. Skewed distribution

Fig. 3.32 Diagrammatic plots of the outputs from two Monte Carlo simulations. The number of answers in successive reserves ranges is plotted against the size ranges themselves. Alternatively one may plot the frequencies as percentages of the total number of answers: the statistical probabilities. Note that the preferred answer that is usually used is the mean value, since it is about this that the standard deviation can be calculated.

this average value, we can work out the standard deviation (the ±) which will give an idea of our confidence in our answer; the smaller the ± range, the more confident we are.

Most usefully, perhaps, we can plot out the percentages of answers in successive size ranges cumulatively: merely add the percentages on to the previous ones as we work up, or rather down, the list (Fig. 3.33). It will give a graph which shows us the chance (probability) that the reserves will be of a certain size or more. This is what is used to determine those reserves

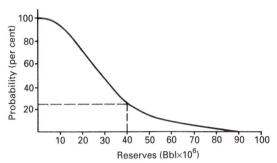

1. Unrisked e.g. 25 per cent chance of finding
   40×10⁶Bbl or more

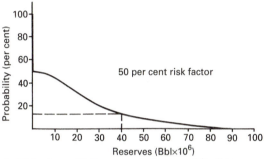

2. Risked e.g. 12.5 per cent chance of finding
   40×10⁶ Bbl or more

Fig. 3.33 The output from a Monte Carlo simulation with the percentages plotted cumulatively. By plotting the answers from the 100 per cent probability downwards, the curve represents the chance (probability) that the reserves are a certain size or greater. In the lower plot, the same values are discounted by a 50 per cent risk factor, to give the chance of discovering certain reserves or more including the 50 per cent chance that we may find nothing at all.

figures that may be called proven, probable, and possible at, say, the 90, 50, and 10 per cent levels of probability respectively, as we mentioned earlier in this chapter. It is also used to assist management in making their exploration/development decisions: for example, if the engineers say that a field of so many million barrels is going to be needed to justify development and production costs, we can read off from the graph the chances of our field containing that much oil or more; management can then decide whether or not to take the gamble on developing the field at those odds. So this type of graph has now become one of the standard key tools in exploration/development decision making.

3.5.2 Undiscovered reserves

This is all very well, you may say, but it assumes that we have already discovered oil; it doesn't take any account of the fact that our exploration well may, for geological reasons, turn out to be totally dry—lacking in hydrocarbons. Indeed it does not! When we are looking at exploration of the unknown, as opposed to assessing what we already know to be there, we have to go a stage further in our crystal ball gazing.

We have to give not only our best estimate of how much petroleum there might be, but also the chance of there in fact being anything there at all. This chance (probability) is known as the *risk factor*: it is an expression, in numbers, of our confidence that there will be at least some oil. The risk factor, combined with the estimate of how much, now gives a more complete picture of the viability or otherwise of an undrilled prospect—at least until we start also considering the costs and economics.

One hears a lot of boloney talked about risk. When it comes down to it, there really is no such thing as *the* risk factor. It cannot be worked out completely objectively, but rather it is the number an individual geologist might produce to reflect his/her personal interpretation of the geology; different geologists will arrive at different figures for the probability of success. And if all this sounds like a gambling game, that is exactly what it is. It is this sort of thing that helps to make the oil exploration business so competitive.

Of course we try to be as scientific, objective, and honest as we can be in assessing exploration risk. The way it is commonly approached is to go back to the 'magic five' (see Section 2.4): remember that all of these essential requirements have to be met if there is to be oil in a particular place and that, if any one of them fails or is lacking, then no oil. We try to assess the probability that each of these five factors will be satisfied, and then merely multiply the five 'conditional probabilities' together to give an overall probability—the risk factor. Simple!

Incidently, one of the main benefits from all of this is that it forces us to think carefully about the geological requirements for there to be oil in our area, and ensures that we consider all possibilities. With the best of intentions, it is all too easy otherwise to get sloppy or carried away: we are only human!

Lastly, on this tack, let us note a number known as the *risked reserves*, the expected reserves estimates from our Monte Carlo simulation multiplied (dis-

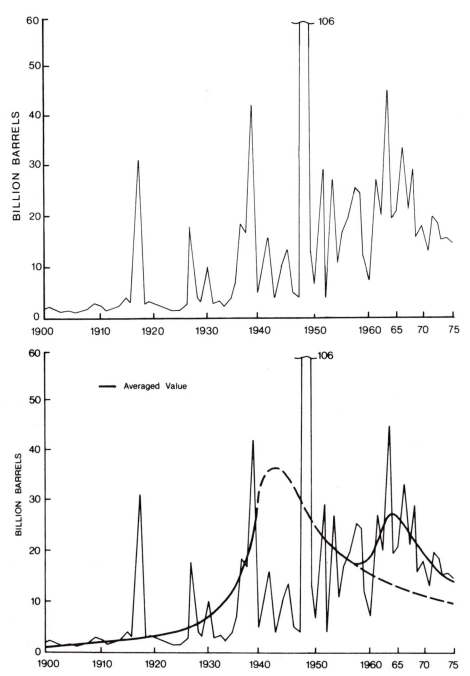

Fig. 3.34 Plot of the annual discovery of oil world-wide since 1900 (after M.T. Halbouty). In the lower diagram, a smoothing line has been applied, averaging out good and poor years; a second hump is added by the fact that, in the 1960s, a new dimension was added to exploration with our ability to work offshore. The significance of this plot is outlined in the text.

counted) by the risk factor (Fig. 3.33). This combines in a single estimate, the two elements of size and chance of success, and as such can be very useful in planning an exploration programme. Should we, for example, go for a large but very risky prospect, or would our money be better spent on drilling a smaller but safer one? The risked reserves, however, is a hypothetical figure, and we should be on our guard against believing that it is what we shall find (it most categorically is not) or otherwise trying to read too much into it.

3.5.3   Ultimate reserves

Now let's have some real fun. What we have been talking about so far, is a single oil accumulation or single prospect. How now do we estimate what still remains to be discovered over a wider area or even an entire sedimentary basin? Out must come the crystal balls again; there really is no precise or objective way of doing it—but still companies and governments want to know.

Many 'experts' have scratched their heads over the estimation of undiscovered reserves, and a number of techniques have been employed. Let us look at the more important.

1. The obvious thing to do is to add together the risked reserves estimates of all the remaining prospects. Some of these will be successful, but some will be dry; the built-in risk factor takes care of this. However, we have to assume that today we can identify and assess all of the prospects that ever will be found in the basin; to believe that we can do this would be the height of conceit.

2. We could adopt what is known as a 'geochemical material balance' approach. This starts with the volume of mature source rock in the basin and then, knowing how rich it is, the amount of oil generated, expelled, and made available for entrapment (the 'charge') can be calculated. There are lots of uncertainties in this but the calculation would be amenable to a Monte Carlo type of simulation. If we have a reasonable amount of information and control, this technique may bring us into the right ball-park; otherwise we may be doing little more than guessing.

3. We might look at explored and known parts of the basin, and calculate average quantities of oil per cubic mile of sediment, or underlying each square mile of surface area; then use these figures for the unexplored parts of the basin.

4. We could make comparisons between known and unknown basins, and use the figures for the known also for the unknown ones.

5. Use past statistics (how many barrels of oil have we found on average for each thousand feet of exploration drilling?) and extrapolate to future drilling. In a similar vein, take a look at Fig. 3.34. It shows the amount of oil found world-wide each year from the beginning of the century; it is a pretty wild sort of plot. However, if we draw a smooth line through it to even out the peaks and the troughs, then the area under it represents the total volume of oil found to date. Extrapolate this smoothing line out into the future, and the area under that bit will represent what, on average, remains to be found. In passing, note that we seem to have already found most of the world's oil, and that what remains for the future doesn't look so rosy! This kind of plot can be used also for individual basins, as well as for the world.

6. If all else fails, get a number of experts to make their forecasts by whatever technique they prefer and, for our 'best estimate', merely use the average of the figures they produce. The approach might be refined by forcing these experts to agree a figure amongst themselves. This is known as the Delphi technique. Delphi was the place in ancient Greece where one went to consult the oracle about one's future; we are said to be consulting the oracles!

How good do you think any of these techniques really is? Which would you prefer? All have been used, sometimes in combination, and some may be more appropriate in given circumstances than the others. But we have to admit that, unless we really have a lot of information (we never have enough!), all of them are very dodgy. Can you, dear reader, think of a better way of knowing the unknown and unknowable? If so, please tell the author and we will go into business together—and make fortunes for ourselves.

4

Reservoir geology

4.1 GENERAL

What happens after a discovery has been made and evaluated is, strictly speaking, beyond the title of this book. However, the geologist and geophysicist will continue to be very much involved during the appraisal, development, and production of a field, and it may be of interest to include a brief chapter.

How this work is organized, and indeed what it is called, vary from company to company. One may encounter the terms 'development geology' or 'production geology' but, since what is involved is essentially a detailed study of the reservoir, 'reservoir geology' may be more all-embracing and less emotive. Much of the work is now in the field of petroleum or reservoir engineering, but the geology still forms the basis of it. Geologists and engineers will therefore be working together throughout as members of a team, and they need to be able to understand and communicate with each other. This in fact often poses a problem, since geologists and engineers have very different backgrounds and use different languages: engineer-speak may not be at all easy for the geologist to follow, or vice versa!

The four main objectives at this stage are:

1. *To understand the distribution of the petroleum in the subsurface.* Not only do we have to try to evaluate the amounts of oil present (refer to Section 3.5), but also how they may underlie individual licence areas. Many accumulations extend under more than one licence block, and therefore will be partly owned by different companies. The proportion of the field belonging to each company is determined by the volume of oil in place beneath its block. It would be uneconomic and generally wasteful for each company to develop and produce its own share separately, as used to happen in the bad old days (Fig. 4.1), and so a field is normally unitized. One company is designated as operator to carry out a programme on behalf of and

agreed by all; the others contribute to development and production costs, and receive a share of the production, in proportion to the oil beneath their land. This business of equity determination, and reaching agreement between the companies, carries a stage further the problems that we have encountered in assessing the field reserves; it really puts the geologist on his mettle, as he/she has to argue with other geologists each trying to obtain for his/her company as large a share of the field as possible.

2. *The location and monitoring of appraisal and production wells.* When a discovery has been made, decisions have to be taken about the optimum number of production wells, and their best locations. This usually requires more information than is obtained from the exploration discovery well. So a few wells are drilled with the primary purpose of obtaining that information: there may be up to four or five of these appraisal wells for a large field, and they may or may not be used later for production.

We do not want to drill more production wells than are really needed. They therefore have to be located so as to obtain the maximum production for as long as possible. In the case of an oilfield, we must delay as far as we can the time when a well will start to produce either water from below the oil–water contact, or gas from above the gas–oil contact. During the course of production, one or other of these contacts, or both, will move (see below), and so the oil production wells should be positioned to try to catch the last of the oil as it is depleted (Fig. 4.2). We have to try to anticipate what will happen in the reservoir during production, in order to avoid wasting money by drilling in the wrong places.

3. *The control of production.* We cannot allow the oil to flow indiscriminately from the reservoir into the well, but must control the rates and levels from which we allow it to come. This is because reservoir forma-

Fig. 4.1 The way things used to be, especially in California and Texas, before drilling was regulated in the USA. In North America, each landowner also owns the underlying oil; each would drill and produce as much and as quickly as possible, resulting in wasted effort, inefficient production, and environmental devastation. Today things are done differently. Photograph by Royal Dutch Shell.

tions are seldom uniform; in some layers the permeability will be high, but in others it will be low or non-existent. Oil will flow preferentially from the high permeability beds, and channelling or perhaps coning may occur (see below). This can result in early water or gas production, and a lot of the oil may be by-passed irretrievably in the reservoir. Detailed knowledge of the geology of the reservoir is required in order to try to minimize these problems.

4. *The control of enhanced oil recovery (EOR) programmes*. It is normally prudent to assist nature in getting the oil to flow through the reservoir and into our production wells by artificial means (see below). Such schemes can significantly increase the percent-

age of the oil in place that we can recover—clearly a desirable objective. Their implementation again demands detailed knowledge of the geology.

4.2 RESERVOIR MODELLING

We aim to accomplish these tasks through the use of a conceptual model of the reservoir in the trap. This model will be pretty primitive just after we have drilled the discovery well, but it will become more and more refined as we get information from additional wells and from more detailed seismic. We shall attempt to model firstly the external shape of the reservoir, i.e. the structure of the top and bottom of the formation, and of

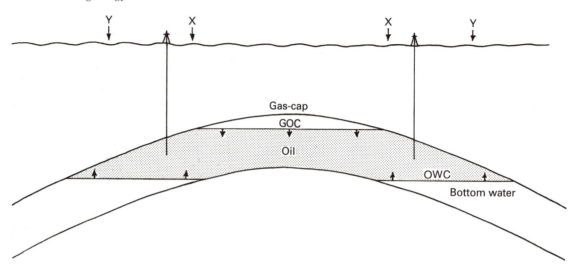

Fig. 4.2 The optimum positions of oil producing wells, assuming that both the gas–oil contact and the oil–water contact will move during production. Wells located at X would quickly go over to producing gas, whereas those at Y would soon produce only water. Wells located as shown may be expected to continued to recover the last of the producible oil.

course its changes in thickness across the field. Then we shall try to incorporate the variations within the reservoir (primarily being concerned with porosity and permeability) into the outline framework.

Once we have produced our geological model, which will have to be adapted to each new piece of information as we get it, the reservoir engineer will take over. He will use it to try to predict how the gas, oil, and water would move through the reservoir during production for a series of different well location patterns and hypothetical production rates. He does this by subdividing the reservoir into a number of 'cells', or blocks, to each of which he can assign values for such parameters as porosity, water saturation, permeability, pressure, etc. This is then used to simulate the anticipated movements of the fluids through the field's production life. The model is essentially a computer-based mathematical one (indeed, it is inconceivable that these games could be played without a large computer), but it is still essentially rooted in the geology.

The geologist will plot out the *external shape* of the reservoir principally from seismic, initially perhaps conventional but later superceded by 3D. Because of the physical properties of the rocks, it is usually the top of the reservoir that can be detected and delineated by seismic; the base may have to be mapped to start with merely by adding on the thickness found in the discovery well. As we get more wells and more refined seismic control, however, the simple initial picture will

be progressively improved. Faults, for example, will be positioned with increasing accuracy.

Mapping the *variations in quality* within the reservoir poses a more serious challenge. We may still get some assistance from seismic, using a technique of lowering the geophones into the well and recording at a number of fixed depths, to give the so-called 'vertical seismic profile' (VSP); or by having the geophones down one well and the source in another, so that the seismic energy travels directly through the reservoir. But most of our data will come from the correlation and interpretation of well logs, and from what we can learn directly from cores.

A commonly used method is to subdivide the reservoir into a number of layers or units, using the wireline logs from our wells for detailed correlation (Fig. 3.31): this may be backed up by micropalaeontology (or, to give the current term, *biostratigraphy*). For each layer, a palaeogeography is constructed in as much detail as possible and, from their inferred environments of deposition, the extent and quality of individual rock units are predicted. For example, if we are dealing with a sequence of deltaic deposits, for each layer we shall try to reconstruct the pattern of the delta, and to identify our sand bodies as channel sands, coastal beaches, offshore sand bars, etc. Sands from each of these environments will have predictable distributions and extents. Or, if we are looking at a carbonate reservoir, we shall try similarly to map out ancient reefs, lagoons, and fore-reefs to forecast the porosity trends.

4.3 PRODUCTION MECHANISMS

When oil is produced from a reservoir, the pore-spaces are not left empty. Very seldom does the ground above collapse to fill a void. Another fluid moves in to take the place of the oil, and this may be associated gas or it may be water. The ways in which the three fluids move, or are moved, in the reservoir environment will affect the flow of oil into our well. This, in turn, governs the production rates and the amounts that can be recovered.

Oil is forced into the well by the pressures of the fluids in the reservoir. If the pressures are sufficiently high, then the oil will flow naturally up the well to the surface: all we shall need is a *well-head* with valves to control the flow (Fig. 4.3), and means of reducing the pressure and separating off the gas that bubbles out of solution (see Section 2.1.1). If the pressure is too low to lift the oil to the surface, then it will have to be pumped (Fig. 4.4).

The removal of oil will tend to reduce the reservoir pressure, and hence retard production, unless it is

Fig 4.3 A flowing production well-head at Wytch Farm, southern England. The valves on the 'Christmas tree' can be adjusted to control the rate of production. Photograph by British Petroleum.

maintained either by natural or by artificial (EOR) means. The types of natural flow through the reservoir are known as *production* or *drive mechanisms*; which one, or more than one, operates is again decided by the geology.

In a *water drive*, the bottom water moves up from beneath as the oil is depleted, pushing the oil–water contact upwards (Fig. 4.5(a)). This will only happen if the reservoir is regionally extensive, perhaps up to the surface; replenishment and minor expansion will ensure a plentiful supply of water under pressure to replace the produced oil. It is the most efficient of the drive mechanisms.

In a *solution gas drive*, gas bubbling out of the reservoired oil as the pressure starts to decrease actually helps to maintain that pressure, and to force the oil out into the well. This process is taken further in a *gas cap expansion drive* (Fig. 4.5(b)). Some of the solution gas will percolate upwards through the oil to join, or even to create, a gas cap which will progressively expand. This pushes the gas–oil contact downwards and again helps to force the oil through the reservoir into the well. This tends to occur if the reservoir is limited in extent and there is not, therefore, a good supply of water.

If the formation is extremely permeable or fractured, a *gravity drainage drive* may cause oil from the upper part of the reservoir to flow under gravity downwards and into the well.

During the course of production of an accumulation, some nasties can arise which we have to try to prevent. In a water (or gas cap expansion) drive, the movement of the oil–water contact (or the gas–oil contact) is seldom even and uniform. The water (or gas) tends to flow preferentially up (or down) the most permeable zones of the reservoir; if uncontrolled, this may result in the water (or gas) being drawn along those zones right through to a well producing from the middle of the oil column. The effect is known as *channelling* (Fig. 4.6). It will mean that we start to produce water (or gas) earlier than need be, and may lead to a lot of oil being left in place in the tighter layers of the reservoir, unproduced and unproducible. The problem is minimized by controlling the rates of production, and the parts of the reservoir that we allow the oil to flow from. This is done by *perforating*, at the desired depths, the steel casing cemented into the well, literally by shooting bullets through it, to make holes to allow the oil to enter the well.

In a similar vein, and if the reservoir is more uniform, we shall of course initially be producing the

Fig. 4.4 Pumps, or 'nodding donkeys', on wells at Wytch Farm, southern England. An electric motor turns the counter-balance on the left, which pushes the back of the beam up and down. Photograph by British Petroleum.

oil that is close to the well. If we are greedy, and try to produce it faster than the oil can make its way through the reservoir from the more distant parts of the accumulation, then the bottom water may be sucked up, or the cap gas from above may be sucked down, close to the well itself. This is referred to as *coning* (Fig. 4.7). It again can mean that we produce water, or gas, early, and again perhaps leave a lot of oil behind.

Clearly, it is a detailed knowledge of the geology of the oilfield, in particular the distribution of permeability, that is required to engineer and regulate production so as to ensure the maximum recovery and the highest efficient flow rates.

4.4   ENHANCED (ASSISTED) OIL RECOVERY

Except in the case of a field with a strong water drive, the production of petroleum will tend to lower the reservoir pressure. This will result in decreasing production rates, and in leaving behind some of the oil that might be recovered. Secondary recovery techniques are designed to help nature in overcoming both problems.

Reservoir pressure is maintained by pumping fluids under high pressure into the formation. Depending on the natural drive mechanism, either water can be injected beneath the oil column or gas can be pumped into the gas cap (Fig. 4.8), to give a boost to the natural flow through the reservoir. The additional quantities of oil that may be recovered from a large accumulation by such techniques may be greater than those discovered in a number of new, small fields. Clearly this is a most desirable aim, and nowadays one or other form of assisted recovery is common practice from the start of production. Note also that gas injection is a form of good housekeeping. Unwanted and useless gas at surface, derived from original solution in the produced oil, instead of just being flared to get rid of it, can be reinjected to help oil production. In this

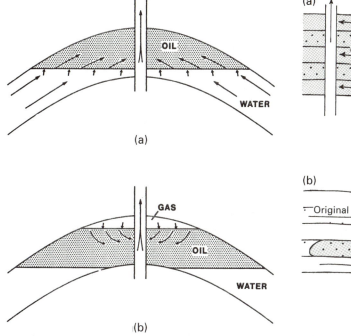

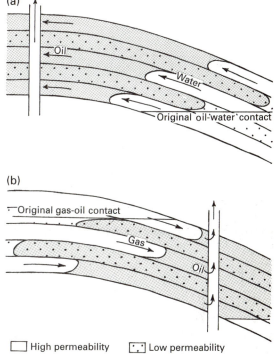

Fig. 4.5 Diagrams illustrating the two principal drive mechanisms acting on an oil accumulation. (a) Water drive; (b) gas cap expansion drive. Please refer to the text for explanation.

Fig. 4.6 Diagrams illustrating the effects of channelling during oil production. (a) Water channelling; (b) gas channelling. Please refer to the text for explanation.

way, it is also being stored for the future: it can be recovered later if the field is converted to gas production when the oil flow has ceased.

Another commonly adopted procedure is to try to improve the permeability of the reservoir in the vicinity of the well. The problem can be compared with traffic coming into London: some distance away it flows freely along several roads but, as we get nearer to the centre, it becomes congested on fewer roads. The similar problem of oil converging on the well may be eased by increasing the permeability either by forcing high pressure fluid into the reservoir to fracture the rock artificially (known as *fraccing*), or by pumping in acid to dissolve away some of the mineral constituents (*acidizing*). The latter is particularly effective in carbonate reservoirs.

Some further, more exotic, techniques are grouped as 'tertiary recovery'. They are, however, expensive and at present generally uneconomic; they are still subject to research. They include pumping detergents into the reservoir to loosen the oil held back in some of the pore spaces (a micellar flood)—a good idea, but the detergents may cost more than the value of the recovered crude oil! Steam injection has been used similarly, particularly to recover heavy viscous oils. And experiments have even been carried out to maintain a slow combustion in the reservoir, to loosen the oil and to drive it towards a production well.

As our known reserves of oil dwindle, it is becoming ever more important to increase the recovery from the accumulations that we have. This is a direction of major research and experiment, to prolong the production of our oil resource into the future—but the implemention and justification of techniques to do so are essentially governed by the price of oil. Much of our business comes back to money in the end.

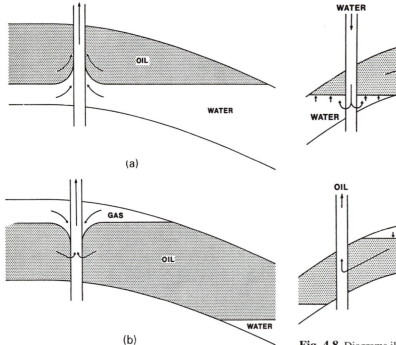

(a)

(b)

Fig. 4.7 Diagrams illustrating the effects of coning during oil production. (a) Water coning; (b) gas coning. Please refer to the text for explanation.

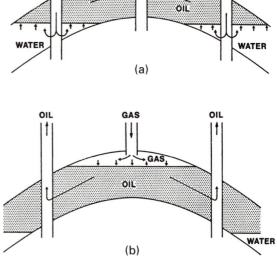

(a)

(b)

Fig. 4.8 Diagrams illustrating the principles of injection for enhanced oil recovery purposes. (a) Water injection; (b) gas injection. Please refer to the text for explanation.

5

North Sea case history

At the beginning of the 1960s virtually nothing was known about the geology of the North Sea. In 30 years it has become a well explored and prolific oil producing region, and the geology is known in considerable detail. This is a remarkable success story of exploration, technological development, and of cooperation between the petroleum industry and national governments. Despite a few sad accidents, it has been achieved with relative safety under extremely hazardous conditions, and with very little effect on the environment. Let us look at the story in the light of what has been covered in this book.

5.1 HISTORICAL PERSPECTIVE

There was little reason in 1960 for the industry to be particularly interested in the North Sea. Small oil and gas fields had been discovered over the preceding decades in the eastern UK, the Netherlands, and northern Germany, but they could not have been economic under more expensive operating conditions offshore. There was no agreement on the ownership of the continental shelf, and no legislative framework for controlling exploration and development. Offshore position fixing and seismic techniques were only beginning to be sufficiently accurate for such work, and the equipment for offshore drilling was adequate only for relatively shallow and calm water operations.

The first 'shot in the arm' came from the discovery of the large Groningen gas field in north Holland in 1959, although it was some 4 years before its vast size, some 95 tcf (trillion or 10^{12} cubic feet), was fully realized. The reservoir was desert sandstones of Permian age, known as the Rotliegend in Germany, the gas being sourced by the coals of the underlying Upper Carboniferous; the seal was provided by Upper Permian evaporites (Zechstein). It was known that similar rocks were present in eastern England, being exposed at outcrop in County Durham: might they

extend at depth across the southern North Sea between the two areas?

By 1964 the 1958 Geneva Continental Shelf Convention had been ratified by the countries surrounding the North Sea and median line agreements were then reached. The way was clear for licencing and legislative procedures, which were quickly enacted. Taxation policies moreover, although naturally varying over the years, have generally encouraged exploration.

Licencing throughout the region was, and is, based on quadrants of 1° latitude by 1° longitude, although the way in which they were subdivided differed in each country (Fig. 5.1). In the UK, for example, there are 30 numbered blocks to the quadrant, and these form the basic exploration licence areas. Licences awarded to individual companies or groups of companies contained seismic and drilling obligations, as well as whole or partial relinquishment requirements. Many relinquished areas have been relicenced subsequently as part blocks. We may therefore see a licence designated as 22/15b: quadrant 22, block 15, relicenced part b. Wells are numbered consecutively within the licence block: at a later stage, oil and gas fields are named by the operating companies. In the other countries the system is similar, although the blocks differ in size between them.

Progress has been rapid since 1964. Early seismic led in 1965 to the large gas discoveries of West Sole (BP) and Leman (Shell/Esso). Numerous further discoveries followed and are still (in 1993) being made, naturally smaller with time but enough to keep the UK supplied with natural gas until far into the twenty first century.

It was soon realized that, to the north, geological conditions were different and that a vast sedimentary basin extended along the centre of the North Sea as far north as the edge of the continental shelf beyond the Shetland Islands; a branch led off westwards into the Moray Firth. Initial seismic and drilling led to a few small discoveries of oil and wet gas, but the real

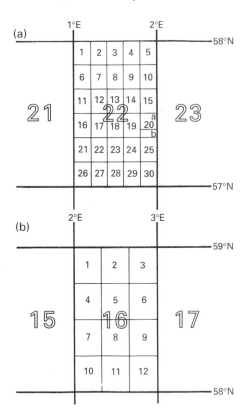

Fig. 5.1 The subdivision of 1° quadrants into licencing blocks in offshore areas of (a) the UK and (b) Norway. Note that the Norwegian blocks are more than twice the size of the British ones. As a result of partial relinquishment and relicencing, blocks may be subdivided, as for example 22/20a and b.

stimulus was provided in 1969 by the large oil find by Phillips at Ekofisk in Norwegian waters. Together with some smaller but economic finds, two major discoveries gave rise to almost a frenzy of exploration; they were Forties in Tertiary sands by BP in 1970 in the central North Sea, and Brent in the Jurassic by Shell/Esso in 1971 in the far north. This was the heyday of exciting exploration and discovery; it was not long before the industry had reached a good general understanding of the geology and of the distribution of the various petroleum plays (see below).

The advance of technology, particularly in seismic, drilling, and production, allowed the industry to carry out detailed and timely exploration under very hostile operating conditions, and has resulted in an unusually high rate of drilling success and the equally unusually rapid development of major production. Although somewhat modest by Middle East and Siberian standards, the North Sea has become a major oil (estimated some 35 billion barrels ultimately recoverable) and gas (approximately 180 trillion cubic feet) province, and smaller but still significant additions continue to be made.

5.2   GENERAL GEOLOGY

We noted in Section 2.4.2 that the North Sea can be classed as a 'failed rift' basin, one in which the 'rift–drift' sequence started to develop by splitting a continent but not so far as to form a new ocean, became aborted, and then continued to subside as a broad saucer as the region cooled down.

Following the creation of a mountain range (the Caledonian) across Scotland and Norway at the end of the Silurian, some 410 million years ago, most of northern Europe was more or less stable through the Devonian, Carboniferous, and into the Permian when our rifting started. During this time, the land was first covered by a desert (Devonian or Old Red Sandstone), then by shallow seas (early Carboniferous), vast tropical swamps in which a lot of coal was formed (late Carboniferous), and then again desert conditions (Permian lasting into the Triassic and forming the so-called New Red Sandstone). This sequence of sediments provided the setting, that we noted above, for forming the gas fields of the southern North Sea, in a region that was not affected as seriously as further north by the later rifting.

Rifting continued through the Triassic, the Jurassic, when the sea flooded in, and the early part of the Cretaceous. A deep central fault-bounded trough was formed, known in the central North Sea as the Central Graben, in the north as the Viking Graben, with the Witch Ground Graben leading off westwards in towards the Moray Firth (Fig. 5.2). It was in these grabens that the thick shales were deposited late in the Jurassic that provide the dominant oil source rocks of the North Sea; sands accumulating around the graben margins provide some of the reservoirs.

In the middle of the Cretaceous, the rifting and most of the associated faulting ceased, and basin-wide gentle subsidence ensued. During the late Cretaceous, sea-level was very high, most of northern Europe was submerged, and there was little land to provide siliciclastic sediment. As a result, the region was blanketed by the very fine-grained white limestone, referred to as

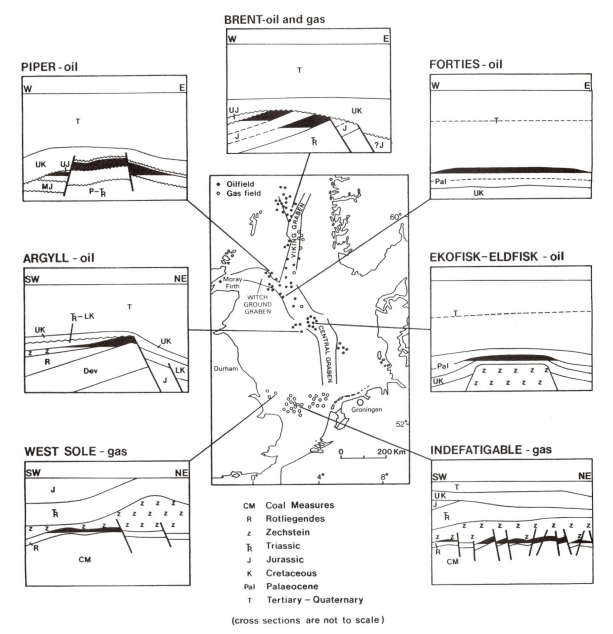

Fig. 5.2 Generalized map of the locations of major North Sea oil and gas fields, together with diagrammatic sections of fields representing the principal plays.

the Chalk and well-known in the white cliffs of Dover. Clastic sediment was again poured into the basin during the Tertiary, mostly derived from the elevated regions of Scotland and Norway; some important reservoir sands spread across parts of the basin. Despite the extension across most of the North Sea of vast glaciers during the Ice Age (Pleistocene), this is the general setting that persists today.

5.3   NORTH SEA PETROLEUM PLAYS

A *play* is defined as a group of fields or prospects in the same region and controlled by the same set of geological circumstances. Thus, for example, most of the gas fields in the southern North Sea have the same Permian sandstone reservoir, the same Upper Carboniferous coal source, are in the same type of trap, and are sealed by the same Permian evaporites: we might refer to, say, the southern North Sea gas play, or the Permian (Rotliegend) gas play. The concept of a play offers a useful framework for looking at the distribution of petroleum in the North Sea. The major plays are summarized here and illustrated by representative fields in Fig. 5.2, but some others also make minor contributions to the total reserves. Note that, in describing them, we are once again coming back to the 'magic five' (Section 2.4).

5.3.1   Southern North Sea gas play

A belt of dry gas fields, some of them large, extends south-east from the Yorkshire coast of England, past Norfolk, and then swings eastwards across northern Holland and Germany.

The source of the gas is the coals in the Upper Carboniferous, the coals that are mined in eastern England and Germany (Fig. 5.3). These source beds are separated from the overlying reservoir by an unconformity. This is important because it provides a first control on the location of the gas fields; they occur where the part of the Carboniferous that actually contains the coals immediately underlies the unconformity, so that the gas can escape upwards into the reservoir.

The reservoir is the Permian desert sands (Rotliegend): the most porous and permeable are those that formed as sand dunes. Sediments that were deposited from flash floods or in temporary lakes have

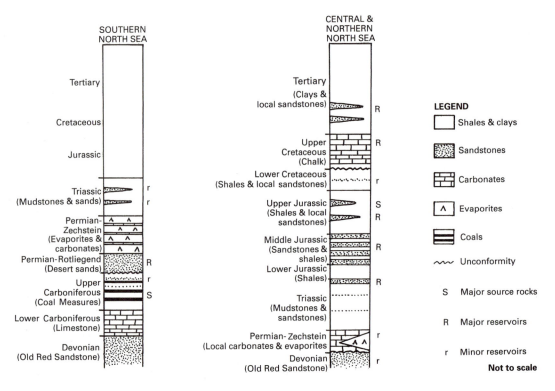

Fig. 5.3 Diagrammatic representation of the successions of strata in the southern and central/northern North Sea basins. The main reservoir and source formations are indicated.

very much poorer reservoir properties, if any. Here, then, is a second control on the distribution of the gas fields: they are found along a belt of former dunes that lay between higher ground to the south and a desert lake to the north.

A third control is provided by the distribution of the Upper Permian salt deposits which form the cap rock or seal. These were precipitated from a supersaline sea that spread over the former sand dunes and covered much of the southern North Sea area. In places, however, the evaporites have been thinned and breached, so that the gas could escape to higher levels; there are a few fields in which gas is reservoired in sandstones of Triassic age.

The traps are rather low structural highs formed principally by faults which do not cut upwards through the plastic salt: two such are shown on Fig. 5.2. The larger horsts and fault blocks have all been drilled, but some of the smaller ones are still (1993) being explored.

We may note also that since the late 1980s exploration has in part been directed to the poorer, tighter reservoir sands interbedded with the coals in the Upper Carboniferous itself. As might be expected, the gas accumulations are smaller and more marginally economic than in the Rotliegend.

5.3.2 Jurassic Brent Sand play

A number of important oilfields lie in the middle of the northern North Sea, between the UK Shetland Islands and Norway; they occur on both sides of the median line between the two countries, and on both sides of the deeply buried Viking Graben.

In middle Jurassic times, there was a pause in the rifting and a river flowed northwards down the centre of the basin and formed an extensive delta in the vicinity of these oilfields. The sands of this delta comprise the reservoirs of this play and are named after the field where they were first found. They were deposited along the shifting beach and in distributary river channels, as well as some other less important environments. Because of the varied local environments of the sands, the reservoirs also are somewhat variable in quality and cause more problems for the reservoir engineers than, say, the Rotliegend reservoir to the south. The seals are provided by the interbedded shales that accumulated in other parts of the delta which shifted their location from time to time.

The source of the oil is a thick marine shale, known as the Kimmeridge Clay (Fig. 5.3) in the UK and the Draupne in Norway, which was deposited most extensively in the Viking Graben. It is essentially in the graben also that it has been buried deeply enough to reach the maturity required for oil generation. The oil, then, has been squeezed out of the Kimmeridge Clay and has migrated up faults into reservoirs actually older, but raised to shallower depth by the faulting.

The traps are formed by the tilting of blocks between sloping faults, the top edges of which were exposed to erosion and later covered, above an unconformity, by Cretaceous shales, which also provide part of the seal (Fig. 5.2). Combination traps of this kind are found also further south along the flanks of the Viking and Central Grabens: the Auk and Argyll fields are cases in point, but there the reservoirs are provided by older, Permian beds.

5.3.3 Upper Jurassic sand play

Sands shed off the uplifted surroundings of the rift basin during the late Jurassic form the reservoirs in a rather diverse set of fields near the margins of the Viking, Central, and Witch Ground Grabens. In almost all cases, the more or less contemporaneous Kimmeridge Clay is the source of the oil (Fig. 5.3).

In Piper in UK and Troll in Norwegian waters, the sands were actually deposited under shallow seas during a slow-down of the rifting; the traps are provided by tilted fault blocks analogous to those containing the Brent fields. In Brae and Fulmar, on the other hand, the sands accumulated against and spread out from the base of cliffs formed by active faults; the traps are in part stratigraphic, being formed by the dying out of the sand accumulations away from the fault scarps.

5.3.4 Upper Cretaceous–Lower Tertiary Chalk play

An important group of fields, in southern Norwegian and Danish waters, overlies parts of the Central Graben in the infill sequence deposited after the rifting of the basin had ceased: Ekofisk may be considered the type example (Fig. 5.2).

Again the source of the oil is the Kimmeridge Clay where it is mature in the Central Graben.

The reservoir in this play is most unusual. It is the extremely fine-grained limestone known as the Chalk (Fig. 5.3). The rock is so fine, and the pore spaces so small, that permeability is normally very low, especially where the irreducible water saturation occupies most or all of the porosity (see Section 2.2.3). In these fields, however, the permeability is abnormally high; there are at least two reasons for this. Firstly, the traps are domes pushed up above the tops of salt plugs (Section 2.3.2), which have severely fractured the Chalk (the oil can move through these fractures). Secondly the pressures in the reservoirs are very high: they are transmitted up from the underlying Kimmeridge Clay which is over-pressured (Section 2.2.1), and they help to drive the oil through the reservoir into the wells so that respectable production rates are obtained.

This play is thus restricted to the vicinity of the Central Graben, where the Kimmeridge Clay is over-pressured, and to those parts of it where there is a sufficient thickness of Upper Permian (Zechstein) salt to form the plugs.

5.3.5   Lower Tertiary sand play

The last of the major plays is in Paleocene and Eocene submarine sands shed into the basin from the uplands of Scotland (Fig. 5.3). The oilfields are located mainly on the western side of the basin in UK waters.

The sand was brought into the basin by rivers, of which the largest flowed down the Moray Firth, and was carried out into the deeper sea beyond the shore-line as a series of fan-shaped spreads by *turbidity currents*. These are suspensions in the water of mud, silt, and sand which travel under gravity, often for long distances, along the sea floor. The beds formed by the sediment that drops out of them are referred to as *turbidites*. Successive such fans covered somewhat different areas, and a reservoir section may be formed by one or more of them. Porosities and permeabilities are commonly high and give good production rates, as for example in the Forties field.

The source of the oil is once again the Kimmeridge Clay, and it must have migrated up into the Tertiary through faults or fractures smaller than the resolution of our seismic. Understanding of this problem has not yet been fully reached.

The traps are largely stratigraphic, created by the dying out laterally of the sand spreads. In some instances, however, there is also structural control; Forties, for example, is in a very gentle drape-compaction anticline (Section 2.3.2) above the eroded remains of a deeper Jurassic volcano, and others are in domes above salt plugs. But it is primarily the positions of the deep sea fans that control the distribution of the play: they are not always easy to locate and new discoveries are still being made.

Appendix A: Suggestions for further reading

Background interest

Yergin, D. (1991). *The prize*. Simon & Schuster, London, Sydney, New York.

British Petroleum Company (1977). *Our industry petroleum*. The British Petroleum Company, London.

General geology

The number of books on general geology is legion. They range from small popular booklets available from museums, etc. to weighty texts for study in universities and beyond. A brief selection only is quoted here.

Brown, G.C., Hawkesworth, C.J., and Wilson, R.C.L. (1992). *Understanding the Earth*. Cambridge Unversity Press.

Clarkson, E.N.K. (1993). *Invertebrate palaeontology and evolution* (3rd edn). Chapman & Hall, London.

Duff, P.McL.D. (1993). *Holmes' principles of physical geology*. Chapman & Hall, London.

Hallam, A. (1977). *Planet Earth: An encyclopedia of geology*. Elsevier–Phaidon, Oxford.

Lambert, M. (1978). *Fossils*. Kingfisher Books, London.

Nield, E.W. and Tucker, V.C.T. (1985). *Palaeontology—An introduction*. Pergamon Press, Oxford, New York.

Pinna, G. (1990). *The illustrated encyclopedia of fossils*. Facts on File, Oxford.

Press, F. and Siever, R. (1986). *Earth* (4th edn). W.H. Freeman & Co., New York.

Roberts, J.L. (1982). *Introduction to geological maps and structures*. Pergamon Press, Oxford, New York.

Particularly useful may be the various handbooks produced for their courses by the Open University, Milton Keynes.

Petroleum geology

There are relatively few books which cover the field of modern petroleum geology. Most books are more or less specialized and go into considerably more detail than a follow-up to the present volume would merit. The following assume some basic knowledge of geology but should for the most part now be understandable.

Hobson, G.D. and Tiratsoo, E.N. (1981). *Introduction to petroleum geology* (2nd edn). Scientific Press, Beaconsfield.

Hyne, N.J. (1984). *Geology for petroleum exploration, drilling and production*. McGraw-Hill, New York.

Levorsen, A.I. (1967). *Geology of petroleum* (2nd edn). W.H. Freeman, San Francisco. (This book is old, but it is a classic and most of it is still valid.)

North, F.K. (1985). *Petroleum geology*. Allen & Unwin, Boston, London, Sydney.

Perrodon, A. (1983). *Dynamics of oil and gas accumulations*. Elf/Aquitaine, Pau.

Selley, R.C. (1985). *Elements of petroleum geology*. W.H. Freeman, New York.

Shannon, P.M. and Naylor, D. (1989). *Petroleum basin studies*. Graham & Trotman, London, Dordrecht, Boston.

Exploration methods

Again, the majority of the available books are specialized and, indeed, some of those listed below are likely to be more than the merely interested reader needs. Some of the basics may be covered best in general petroleum geology books mentioned above.

Anstey, N.A. (1982). *Simple seismics*. International Human Resources Development Corporation, Boston.

Asquith, G. and Gibson, C. (1982). *Basic well log analysis for geologists*. American Association of Petroleum Geologists, Tulsa.

Drury, S.A. (1987). *Image interpretation in geology*. Allen & Unwin, London, Boston, Sydney.

Griffiths, D.H. and King, R.F. (1981). *Applied geophysics for geologists and engineers* (2nd edn). Pergamon Press, Oxford.

Helander, D.P. (1983). *Fundamentals of formation evaluation*. Oil and Gas Consultants International, Tulsa.

Kearey, P. and Brooks, M. (1984). *An introduction to geophysical exploration*. Blackwell Scientific, Oxford.

McQuillan, R., Bacon, M. and Barclay, W. (1984). *An introduction to seismic interpretation*. Graham & Trotman, London.

Rider, M.H. (1986). *The geological interpretation of well logs*. Blackie, Glasgow, London.

Selley, R.C. (1985). *Elements of petroleum geology*. W.H. Freeman, New York.

Tucker, M.E. (1982). *The field description of sedimentary rocks*. Geological Society of London Handbook, Open University Press, Milton Keynes.

Reserves estimation

Harbaugh, J.W., Doveton, J.H., and Davis, J.C. (1977). *Probability methods in oil exploration*. John Wiley & Sons, New York.

Haun, J.D. (ed.) (1975). *Methods of estimating the volume of undiscovered oil and gas resources*. American Association of Petroleum Geologists, Studies in Geology 1, Tulsa.

McCray, A. W. (1975). *Petroleum evaluation and economic decisions*. Prentice-Hall, New Jersey.

Newendorp, P.D. (1975). *Decision analysis for petroleum exploration*. Petroleum Publishing Company, Tulsa. (Although the emphasis is on the economics, this book is a classic.)

Reservoir geology

Again there are few books that are not too specialized or complex. The following is one of the few suitable:

Dickey, P.A. (1979). *Petroleum development geology*. Petroleum Publishing Company, Tulsa.

North Sea

There is really just one volume that gives an overview, and one small guide that gives a basic introduction. Other publications are detailed, specific, and found in the geological literature.

British Museum (Natural History) (1988). *Britain's offshore oil and gas*. BM(NH), London.

Glennie, K.W. (ed.) (1990). *Introduction to the petroleum geology of the North Sea* (3rd edn). Blackwell Scientific, Oxford.

Appendix B: Glossary of technical terms

Please do not be put off by the length of this list. Not all the terms are needed every day, but they are included for reference purposes.

Accumulation (*pool*). A single concentration of petroleum found in a reservoir in a trap.

Anhydrite. An evaporite mineral formed of calcium sulphate ($CaSO_4$). The related form *gypsum* contains water in the crystal lattice.

Anomaly. Usually used in relation to gravity and magnetic surveys. Measurements that are above (positive) or below (negative) the regional average.

Anoxia. A lack of life-giving oxygen. Usually refers to conditions at the sea-bottom under which source rocks accumulate.

Anticline. An upfold of strata.

API gravity. A measure of the density of oil, in degrees, on a scale devised by the American Petroleum Institute. API gravity is inversely proportional to specific gravity.

Aromatics. A family of hydrocarbons, in which the basic structure is a ring of six carbon atoms with single and double bonding.

Associated gas. Gas occurring together with oil, either dissolved in it or as a separate gas cap.

Barrel. The unit commonly used for measuring volumes of oil; 1 barrel = 42 US gallons or 35 Imperial gallons.

Basement. An informal term used to refer to the rocks underlying the sediments of interest for petroleum. Usually igneous or metamorphic rocks.

Bedding. The natural layering of sedimentary rocks.

Biogenic gas. Mainly methane produced by the bacterial decay of organic matter in muds shortly after their deposition.

Bit. The cutting device rotated on the bottom of a well during drilling to chip away the rock and deepen the well.

Blow-out. A dangerous uncontrolled flow of high pressure fluids up a well during the course of drilling, either drilling mud, water, gas, or oil. Normally controlled by *blow-out preventers* (BOPs).

Calcite. A calcium carbonate ($CaCO_3$) mineral which is the common constituent of limestones. Forms some shells, corals, etc.

Cap rock (*seal*). An impermeable layer overlying a reservoir formation. Required to retain petroleum within the reservoir. In America normally refers to the leached top of a salt plug.

Carbonate. A blanket term to include limestones and dolomites.

Casing. Steel pipe cemented into a well after a section of it has been drilled, to protect the well bore from collapse. It also holds back fluids in the rocks.

Channelling. The preferential movement of oil, and gas or water, along the most permeable layers of a reservoir formation. Can result in unwanted early production of gas or water.

Charge. The volume of petroleum available to fill a trap, as a result of generation and migration.

Clastic sediments. Sediments composed of particles derived from pre-existing rocks by erosion, fragmentation, and transportation.

Closure. The vertical height of a trap between the crest and the spill-point. Also refers to the area of the trap enclosed above the spill-point.

Coal. A combustible rock formed by the compaction, induration, and maturation of plant material. The *rank* reflects the degree of thermal alteration (maturity).

Column. The vertical thickness of a petroleum accumulation.

Combination trap. A petroleum trap formed partly by structural and partly by stratigraphic effects.

Common mid point (CMP). A point on a rock layer from which more than one reflection is obtained at surface during a seismic survey. The records of the different reflections can be added together to enhance the real signal and to cancel out background noise.

Compaction. The reduction in volume of a sediment as water is squeezed out of it during burial. Normally applied principally to muds and shales.

Compression. Normally refers to the horizontal squeezing and folding of rocks or strata by forces acting within the Earth.

Condensate. Petroleum occurring as gas under subsurface conditions, but as liquid at the surface.

Conglomerate. A coarse clastic sedimentary rock formed by the induration of shingle.

Coning. The drawing up of water from below, or the drawing down of gas from the gas cap, close to the well during oil production from a reservoir. Can result in unwanted early production of water or gas.

Contact. The boundary between accumulations of different fluids in a reservoir. Thus, gas–oil contact, oil–water contact, gas–water contact.

Continental drift. Movements of the continents relative to each other as a result of plate tectonics.

Core. A vertical cylinder of rock cut during the drilling of a well to provide a large sample for description and analysis.

Correlation. The matching up and comparison of the successions of strata drilled by different wells or seen at outcrop.

Cracking. The breaking down, usually by heating, of large hydrocarbon molecules into smaller ones, of heavy oils into lighter ones.

Crest. The highest point of a trap.

Cross-section. A representation of strata and their attitudes, as an imaginary vertical slice through the Earth looked at sideways. Normally constructed to scale.

Cuttings. The small chips of rock cut by the bit during the drilling of a well.

Derrick floor. The elevated platform at the foot of the derrick on a drilling rig, from which the drilling operations are controlled.

Diagenesis. The alteration, by various processes, of sediments and sedimentary rocks after their original deposition. Also refers to the first stage of maturation of a source rock, during which biogenic gas may be generated.

Diapir. A salt plug.

Dip. The angle of slope, measured from the horizontal, of the bedding of sedimentary rocks, or of a fault plane.

Direct hydrocarbon indicator (DHI). An indication on a seismic profile of the presence of petroleum, usually gas. DHIs include *flat spots*, reflections from a gas–liquid contact, and *bright spots*, enhanced reflections due to the presence of gas as opposed to water or oil in a reservoir.

Dolomite. A rock composed of the calcium–magnesium mineral of the same name. Dolomites may be precipitated directly from sea-water, or may be formed by post-depositional alteration of limestone.

Dome. An anticline approximately circular in plan.

Drape anticline. An anticline formed by the drape of sedimentary layers over the top of a raised feature. Often enhanced by differential compaction.

Draw-works. The massive horizontal winch on a drilling rig which carries the cable used for raising and lowering the drill-string.

Drill-string. The steel pipe carrying the bit and which is rotated in the well during drilling. Composed of 30 ft sections (*singles*), the bottom few of which are particularly heavy (*drill collars*).

Drilling mud. A liquid mud which is pumped down the drill-string, out through nozzles in the bit and up the space between the drill-string and the wall of the well. It cools and lubricates the bit, carries the cuttings to surface and supports the well bore.

Drive mechanism. See Production mechanism.

Dry gas. Natural gas composed almost entirely of methane.

Dry hole. A well that failed to discover any petroleum. Sometimes informally called a 'duster'.

Dyke. A wall of igneous rock intruded into preexisting rocks.

Effective permeability. The permeability of a rock to a particular fluid if more than one fluid is present.

Effective porosity. The porosity of a rock available to be filled by oil or gas. Also defined as the interconnected porosity.

Enhanced oil recovery (EOR). The techniques of increasing petroleum recovery from an accumulation by artificial means.

Erosion. The gradual fragmentation of rocks at the land surface by the actions of water, ice, and wind, and the removal of the resulting debris.

Evaporites. Sedimentary rocks formed by the precipitation of salts from sea-water. They include some limestones and dolomites, anhydrite ($CaSO_4$), rock salt (NaCl), and potassium salts.

Facies. A convenient term referring to some characteristic of rocks; thus 'sandstone facies', 'coarse-grained facies', 'facies distribution', etc.

Fault. A cross-cutting break in the rocks, where one side is moved relative to the other.

Field (oilfield or gasfield). The various accumulations of oil and/or gas lying within a common geological structure.

Formation. A recognizable, describable, and mappable group of related strata. Usually named after a place.

Fossil. The remains of an animal or plant preserved in a sedimentary rock. Usually only the hard parts of the skeleton are preserved.

Gas cap. An accumulation of gas trapped in a reservoir above an oil accumulation.

Geophone. The instrument used in a seismic survey for receiving the seismic energy reflected back to surface. At sea, the corresponding device is known as a *hydrophone*.

Graben (rift). An elongated trough dropped down by normal faults on either side.

Gravimeter. The instrument used for very accurate measurement of the Earth's gravity during a gravity survey.

Gravity survey. A geophysical technique of investigating the subsurface geology through the measurement of small changes in the acceleration due to gravity (*g*) across the survey area.

Halite. Rock salt (NaCl).

Halokinesis. The flowage of salt in a layer in the subsurface: may lead to the formation of salt domes and salt plugs.

Hiatus. An interruption in the continuity of a sequence of sedimentary layers. Also used to refer to the time represented by such an interruption.

Horst. An upraised block between two normal faults. The inverse of a graben.

Hydrocarbons. Chemical substances composed essentially of carbon and hydrogen atoms. The components of petroleum, for which the word is often used interchangeably.

Hydrophone. See Geophone.

Igneous rocks. Rocks formed by the cooling and solidification of molten material (*magma*) from the Earth's interior.

Induration. Hardening, to form rock from loose sediment.

Kelly. The top section of the drill-string, square or hexagonal in cross-section. It slides up and down through a corresponding hole in a rotating *kelly bushing* in turn rotated by the *rotary table*, in order to turn the bit at the bottom of the drill-string. Well depths are measured from the elevation above sea-level of either the kelly bushing or the rotary table.

Kerogen. The initial alteration product of organic matter during maturation of source rocks. The actual substance from which petroleum is generated thermally.

Limestone. A sedimentary rock composed primarily of the calcium carbonate mineral calcite. The calcium carbonate may have been precipitated from sea-water, but more commonly is derived from the secretions or remains of organisms. Limestones are classified according to their constituents, particle or crystal size, and texture.

Lithology. Rock type. The nature and mineral composition of a rock.

Log. A record, usually in graphical form, of the rocks penetrated by a borehole. *Wireline logs* are those that are recorded after the well has been drilled to measure some physical property of the rocks.

Magma. See Igneous rocks.

Magnetic survey. A geophysical technique for investigating the subsurface geology, through the measurement of variations in the Earth's magnetic field across the survey area. Usually conducted from aircraft carrying a sensitive *magnetometer*.

Maturation. The alteration of kerogen in source rocks, initially by bacterial activity and then under the influence of the Earth's internal heat, resulting in the generation of petroleum. A source rock that has been heated sufficiently to generate oil, i.e. which is within the 'oil window', is said to be *mature*.

Measurement while drilling (MWD). The logging of a well during the course of drilling by instruments carried close to the bit.

Metamorphic rocks. Rocks formed by the reconstitution of pre-existing rocks by intense heat and pressure from within the Earth. They may originally have been either igneous or sedimentary.

Methane. The simplest and lightest of the hydrocarbons: the molecule consists of one carbon and four hydrogen atoms. The principal constituent of natural gas, formed either biogenically or thermogenically.

Migration. (a) The movement of petroleum in the sub-surface; we distinguish primary and secondary migration. (b) The application of a correction to seismic records to restore the point of origin of a seismic reflection to its true subsurface position.

Mudlog. A preliminary log of a well compiled during drilling by examination of the cuttings, measurement of the rate of drilling, and analysis of the drilling mud for traces of hydrocarbons.

Mudstone. A clastic sedimentary rock formed by the compaction and induration of clay or mud.

Mud volcano. A small volcano-like feature formed by the drying out of liquid mud, sometimes containing gas, that has escaped to surface from below. Commonly indicative of over-pressures at depth.

Naphthenes. A family of hydrocarbons in which the basic structure is a ring of singly bonded carbon atoms.

Normal fault. A fault in which the downthrown side is in the direction of dip of the fault plane. Produced by horizontal extension.

Oil in place. The total volume of oil in an accumulation in the reservoir. Not all of this amount is recoverable.

Oil shale. A shale containing sufficient kerogen to generate and yield exploitable quantities of oil when heated artificially.

Oil window. The depth range of the time–temperature conditions where oil is generated from kerogen within a source rock. Very approximately between 150 °F and 350 °F. Sometimes referred to informally as the 'oil kitchen'.

Orogeny. The process of creating a range of mountains, commonly by horizontal compression. Used specifically to refer to particular episodes of mountain building.

Overpressures. Subsurface pressures greater than those caused by a column of water extending up to surface. Mainly found in thick mudstones, which have a higher content of water than normal for their depth (*undercompacted*), or in completely sealed sand lenses. Such overpressures may approach the weight of the rock overburden (*geopressures*).

Palaeogeography. A reconstruction of the physical geography of an area at a particular time in the past.

Palaeontology. The study of fossils, their morphology, classification, evolution, and ages.

Palynology. The study of fossil pollen and spores.

Paraffins (alkanes). A family of hydrocarbons in which the basic structure is a chain of bonded carbon atoms.

Pay. The petroleum-bearing interval(s) penetrated by a well.

Perforation. The making of holes at controlled depths through the casing in a well, by shooting bullets through it from a device lowered on a cable from the surface, to allow petroleum to flow into the well.

Permeability. A measure of the ease with which a fluid can pass through a rock. Expressed in *darcies*.

Petroleum. A blanket term to include all naturally occurring oils, gasses, and solid hydrocarbon substances.

Plate tectonics. The concept that segments of the Earth's crust are in continuous motion relative to one another. New oceanic crust is formed at *spreading ridges*, and old crust is destroyed at *subduction zones*. Plates thus have *constructive* and *destructive margins*, as well as *transform margins* where one plate slips laterally past another.

Play. A group of fields and/or prospects in the same general area and having the same overall geological characteristics.

Pool. See Accumulation.

Porosity. The volume of the void space in a rock, expressed as a percentage of the bulk rock volume.

Primary migration. The movement of oil and gas within and out of their source rock.

Production mechanism. The natural process driving oil through and out of the reservoir into production wells; e.g. a water drive, a gas cap expansion drive. Also referred to as a *drive mechanism*.

Recovery factor. The percentage of the oil in place in an accumulation that can be produced to surface.

Reserves. Normally refers to the volume of oil or gas that remains at a specified date to be produced commercially from a field or area under prevailing economic conditions.

Reservoir. A sedimentary layer or formation with sufficient thickness, porosity, and permeability to contain commercial petroleum and to yield it to a producing well. Most commonly a sandstone or carbonate, although occasionally other rock types can act as reservoirs.

Reverse fault. A fault where the upthrown side is in the direction of dip of the fault plane. Produced by horizontal compression.

Risk factor. A statistical probability expressing one's confidence that a prospect or area will prove to contain petroleum.

Roll-over anticline. An anticline formed by the rotation into the fault plane of beds on the downthrown side of an upwards curved (listric) normal fault. Generally associated with faults that were active during deposition of the sediments affected (growth faults).

Salt dome. A dome produced by the upwards bulging of an underlying thick layer of salt.

Salt plug. A body of salt cutting upwards into and through the beds overlying a thick mother layer of salt.

Sandstone. A clastic sedimentary rock formed by the induration of sand. Sandstones are classified according to the size and composition of their constituent grains, and on the nature of the intergranular matrix and cement.

Seal. See Cap rock.

Secondary migration. The movement of oil and gas through reservoir formations or other channels, after their expulsion from the source rock.

Sedimentary basin. A region where thick sedimentary sequences have accumulated. Formed either by crustal subsidence or by the spread of abundant sediment into deep water.

Sedimentary rocks. Rocks formed by the induration of sediment that accumulated originally on the sea-floor, in lakes, rivers, or on land.

Seepage. A natural escape to surface of accumulated oil or gas, where at least the lighter hydrocarbons evaporate and are lost.

Seismic source. An artificial source of the energy that is applied to the ground in a seismic survey.

Seismic stratigraphy. The use of seismic profiles to interpret the subsurface relationships of sedimentary layers and their environments of deposition.

Seismic survey. Usually refers to seismic reflection survey, a geophysical technique to investigate the subsurface geology by measuring, at a series of stations, the time that energy takes to travel from the source, down to a rock layer, and back to a geophone at the surface. The results are processed, compiled, and displayed on a *seismic profile*, which is similar to but different from a cross-section, depths being measured in *two-way travel time* (TWT).

Shale. A very fine-grained rock formed by the induration of mud. Strictly speaking, the rock is fissile along the bedding but the term tends to be used indiscriminately in place of 'mudstone'.

Sidewall sample. A small cylinder of rock cut from the wall of a borehole by a device lowered on a cable from the surface.

Siliciclastic. Refers to the silicious nature of clastic sediments. Silica, in the form of quartz, is the most abundant constituent of sandstones.

Siltstone. A clastic sedimentary rock formed by the induration of silt.

Sonde. A device lowered into a well on a coaxial cable and carrying instrumentation to measure physical properties of the rocks as it is withdrawn. See Wireline logs.

Source rock. A sedimentary rock containing organic matter that has generated, or has the potential to generate, petroleum. Most commonly a shale containing at least 0.5 per cent of kerogen.

Spill point. The lowest point of a trap, i.e. the point at which petroleum would spill out if overfilled. The term is also used for the depth of that point.

Spreading ridge. See Plate tectonics.

STOOIP (stock tank oil originally in place). The volume of oil in an accumulation in a reservoir but corrected to the volume it would occupy at surface temperature and pressure.

Stratigraphy. The study of the nature, succession, and relationships of sedimentary rock layers or *strata* (singular *stratum*).

Strike. The direction of a horizontal line drawn on a sloping surface. It is at right angles to the direction of dip.

Structure. Usually refers to the attitude of the sedimentary layers resulting from folding or faulting. An anticline is a structure which may contain oil.

Structure contour map. A map showing contours drawn on a level of interest in a sedimentary sequence, often the top of a reservoir formation. The contours are in feet, or metres, *below* sea-level.

Subduction zone. See Plate tectonics.

Synthetic seismogram. A simulation of a seismic trace at a well, calculated and plotted from physical properties of the rocks measured by wireline logging. Used to compare the reflections on a seismic profile with rock layers drilled through.

Tar sand. A sand or sandstone at surface impregnated with heavy residual oil, the lighter components of the crude having been lost. The term normally refers to those occurrences that are extensive enough to have the potential for exploitation.

Tectonics. Large-scale deformation of the Earth's crust, commonly leading to folding, faulting, mountain building, graben formation, etc.

Thermogenic gas. Natural gas generated by the thermal maturation of kerogen, especially at high temperatures where oil is no longer being produced.

Trap. A subsurface environment where petroleum migrating upwards through a reservoir is barred from further migration and therefore accumulates. Formed by structural (folding or faulting) or stratigraphic (lateral discontinuity of the reservoir) controls.

Turbidite. A layer of sand and mud transported by and deposited from a turbidity current—a suspension of sediment flowing under gravity along the sea-floor.

Turtle-back (turtle) structure. A residual anticline commonly found between two salt plugs.

Unconformity. A hiatus where there is angular discordance between the beds above and below.

Velocity (seismic). The speed with which seismic energy travels through a rock layer.

Vertical seismic profile (VSP). A seismic record obtained by placing geophones at different depths in a well. The technique gives clear and accurate data close to the well.

Vibroseis. A source of seismic energy using a vibrating plate lowered to the ground surface beneath a heavy truck.

Vitrinite reflectance. A laboratory technique for measuring the maturity of plant material preserved in a source rock. The shinier the material, or the more light that it reflects, the higher the level of maturation.

Water saturation. The percentage of porosity that is occupied by water as opposed to petroleum. The inverse is known as the *oil saturation*.

Wet gas. Natural gas that contains more than a trace of hydrocarbons heavier and more complex than methane, i.e. ethane, propane, and butane. It is usually associated with oil accumulations.

Wildcat. An exploration well drilled on a new prospect or in a new area.

Wireline logs. Records of particular physical properties of the rocks obtained by lowering a sonde down the well on a cable, which transmits a continuous record to surface as the sonde is withdrawn. A suite of such logs recording different properties forms the standard method of logging wells, and can be interpreted to reveal lithology, porosity, and fluid contents.

Index